Building Services Design Methodology

Building Services Design Methodology: A practical guide clearly sets out and defines the building services design process from the feasibility stage through to post-construction design feedback. By providing a simple step-by-step methodology the book describes how compliant and creative design solutions are matched to the clients project requirements encouraging design success and repeat work. The generic advice and guidance is set in the current legal and contractual context making this volume required reading for all building services professionals and students.

David Bownass C.Eng., MSc., B.Eng., MIMechE, MCIBSE, MIOA. is a Technical Director with WSP, the construction services group. He has spent 18 years in Building Services engineering, working for contractors, design consultants and client organisations. His experience ranges from pre-design client liaison work through to post-construction commissioning and final account close out. In recent years he has concentrated on front end project work and the management of Building Services design and installation contracts.

Building Services Design Methodology

A practical guide

David Bownass

London and New York

First published 2001 by Spon Press
11 New Fetter Lane, London EC4P 4EE

Simultaneously published in the USA and Canada
by Spon Press
29 West 35th Street, New York, 10001

Spon Press is an imprint of the Taylor & Francis Group

© 2001 David Bownass

Typeset in Sabon by
Deerpark Publishing Seervices Ltd
Printed and bound in Great Britain by
TJ International Ltd., Padstow, Cornwall

British Library Cataloguing in Publication Data
A catalogue record for this book is available from the British Library

Library of Congress Cataloguing in Publication Data
Bownass, David A. (David Arthur), 1960-
    Building services design methodology: a practical guide / David Bownass.
        p. cm.
    Includes bibliographic references and index.
    ISBN 0-419-25280-0 (pbk. : alk paper)
        1. Buildings--Mechanical equipment--Design adn construction. I. Title.

TH6010 B68 2000
696--dc21

                                                                    00-030902

ISBN 0-419-25280-0

Contents

List of figures

List of tables

Acknowledgements

Crown copyright material is reproduced with the permission of the Controller of Her Majesty's Stationery Office. Extracts of Crown copyright material are drawn from the following texts:

Health & Safety at Work Act, 1974

A Guide to Managing Health and Safety in Construction, Health and Safety Executive, 1995

INDG270: Supplying New Machinery, Health and Safety Executive, 1998

Extract from BS 7000 Part 3: 1994 reproduced with the permission of BSI under licence number 2000SK/0089. Complete standards can be obtained by post from BSI Customer Services, 389 Chiswick High Road, London, W4 4AL.

By-laws 33 and 34 of the Institution of Mechanical Engineers, code of conduct reproduced with kind permission of The Institution of Mechanical Engineers.

By-law 38 of The Chartered Institution of Building Services Engineers, Royal Charter and by-laws reproduced with kind permission of The Chartered Institution of Building Services Engineers.

Abbreviations

ASHRAE	American Society of Heating, Refrigeration and Air Conditioning Engineers
BEMS	building energy management systems
BMS	building management systems
BSRIA	Building Services Research and Information Association
BS	British Standard
BSI	British Standard Institute
BREEMA	building research establishment environmental assessment method
CAD	computer aided design
CE	Conformité Européen
CD ROM	compact disc read only memory
CDM	Construction (Design and Management) Regulations 1994
CHP	combined heat and power
CIBSE	The Chartered Institution of Building Services Engineers Guide
CYMAP	building services design software
°C	degree centigrade temperature scale
DSR	design safety review
dB	decibel
dB(A)	'A' weighted dB sum, usually across 63 Hz to 8 kHz octave bands
ddc	direct digital control
EA	Environmental Agency
EC	European Commission
ECA	European Communities Act 1972
EEMUA	The Engineering Equipment and Materials User Association
EFTA	European Free Trade Association
EHO	environmental health officer
EU	European Union
g/kg	grams per kilogram
GMP	Good Manufacturing Practice
FAST	functional analysis system technique
FDA	Food and Drug Administration
FIDIC	Federation Internationale des Ingenieurs-Conseils
FS	Federal Standard
FM	factory mutual
HASAWA	Health and Safety at Work Act 1974

HAZOP	hazard and operability study
HEPA	high efficiency particle absorption filter
HEVACOMP	building services design software
HSC	Health and Safety Commission
HSE	Health and Safety Executive
HS(G)	health and safety guidance
HTM	health technical memoranda
HMSO	Her Majesty's Stationary Office
HVAC	heating, ventilating and air conditioning
Hz	hertz or cycles per second.
IChemE	Institute of Chemical Engineers
LEL	lower explosion limit
IMechE	Institute of Mechanical Engineers
IOP	Institute of Plumbing
IPPC	integrated pollution prevention and control
ISPE	International Society of Pharmaceutical Engineering
JCT	Joint Contracts Tribunal
kHz	a thousand cycles per second
kW	kilowatts, a unit of energy
kPa	kilopascals, a unit of pressure
LEV	local extract ventilation
LPC	loss Prevention Council
LTHW	low temperature hot water
L_{eqA}	equivalent 'A' weighted noise level
L_{90}	the unweighted noise level exceed 90% of the time.
MDPE	medium density polyethylene
MCA	Medicines Control Agency
m^3/s	cubic metres per second
m/s	metres per second
NHS	National Health Service
NSCA	National Society for Clean Air and Environmental Pollution
NR	noise rating standard
NFPA	National Fire Protection Association
NPSH	net positive suction head
Octave band	a noise sample bandwidth
OEL	occupation exposure level
O&M	operation and maintenance
Pa	Pascal
PC	personal computer
PPE	personal protective equipment
PRV	pressure relief valve
QA	quality assurance
RFI	request for information
RIBA	Royal Institute of British Architects
R.H.	relative humidity
rpm	revolutions per minute
TM	technical memoranda

VAV	variable air volume
VOCs	volatile organic compounds
μ/m^3	micrograms/cubic metre

Preface

The concept for the book developed slowly, probably over a period of 12 months. It started with a feeling that many design engineers were so immersed in their work that they could not see the design process. This lead to concern that the requirements for the different design stages where frequently misunderstood and developed into 'how can the design process be improved?'. Finally the need to find a dissertation subject that satisfied the requirements of Warwick University's engineering business management course crystallised the thoughts into actions. The dissertation took 18 months to complete and a further 2 years to convert into a book.

The book is intended for:

- All students thinking about or studying building services, as it describes the practical process involved rather than the physics of the subject.
- Practicing designers and design managers who may need to balance their technical skills with a better understanding of the design process.
- Clients and project managers who are not familiar with managing or procuring projects that contain significant amounts of building services engineering.

I apologise in advance to all my electrical colleagues as although the design *methodology* applies equally to them the chapters do have a mechanical bias. Subject to the book's reception I will persuade one of them to add the appropriate work in any future edition.

The book is based on my research, opinion and experience. As nobody has complete knowledge of their subject, particularly in a large and complex industry like building services engineering, I anticipate it will generate a lot of comments. In many ways it is a starting point and I welcome your feedback. The quickest way to get in touch with me is via e-mail at david@abownass.freeserve.co.uk

Note to readers

This book is intended to provide an explanation of the building services design process. It is not definitive and readers should excerise their own judgement in deciding whether to apply any of its content.

Legal, professional and technical requirements frequently change to accomodate new technological advances and industry feedback. Readers need to ensure they take appropriate professional advice and use current and up to date information in developing their design solutions.

David Bownass

Chapter 1

Introduction

1.1 Chapter contents

The topics covered in this chapter include:

- The aim of the book
- How to use the book

1.2 The aim of the book

The design of building services systems has changed considerable in the last 20 years. The change in approach, procurement, design tools and legislation has been phenomenal.

The consultant's role has changed from the authoritative adviser to a provider of design services and information. At the same time the industry has tended to segment into large design consultants, small niche design practices and a huge range of specialist subcontractors providing design and installation packages. Designers are no longer simply required to sit at a drawing board, produce designs and advise the client of their needs. There are few, if any drawing boards left in modern design offices! The designer is more likely to be sitting at a computer editing a document or integrating specialist suppliers' packages into the design solution or presenting the design to the client. The designers' skills have expanded from a simple emphasis on technical knowledge to include information technology skills, communication skills, presentation skills, finance and design management.

The procurement route for construction work has also changed. Design and build and its many derivatives now appears to be the main procurement method. The traditional, architect led contract is now the exception rather than the norm.

Design tools have evolved beyond recognition. Drawing boards, manual hand calculations, drawing pens, log tables, carbon copies and typewriters have all been replaced by the computer and information technology. Almost all drawing is now on computer aided design (CAD) stations. Computers are used for most calculations, they provide access to a growing range of information on the internet and act as the main means of exchanging information through e-mail and project document management systems. They can even model the design solution before it is built! Although teleconferencing is still not commonly used within the industry, mobile communications particularly mobile phones are accepted as a standard accessory. The use of letters has declined dramatically. They are only used for official correspondence as a record of events; the fax and increasingly e-mail have superseded them.

Legislation affecting design has increased significantly. The vast majority of this legislation relates to health, safety and the environment. More subtly, the legislation implementation has changed, shifting the emphasis onto the designer. Legislation used to be prescriptive, e.g. guards shall enclose all rotating machine parts. Current legislation is descriptive placing a duty on the designer to use 'a reasonably practicable' approach in defining the solution.

In summary, design services in the early twenty-first century are procured rather than appointed, the designer's skill base has broadened, there are more routes to procuring construction projects, information technology is an integral part of design work and there is more legislation affecting design. All these fundamental changes fit within a business culture that is more competitive, more global and a lot faster; time to market is often critical.

While the changes are significant, the rate of change is likely to accelerate. This is because the construction industry has been relatively slow to grasp the 'service culture' and starts from a point well behind other comparable industries. For instance, new teams of people are formed for virtually every project, financial disputes are still common place, construction productivity remains low and employment can be erratic.

There are a number of business and design issues that will force further change on the industry. The key business issues are global competition, alternative government procurement strategies, e.g. private finance initiative, and the consequences of the Egan Report. The Egan Report is currently being considered by the construction industry and it is likely to have a significant impact on design and the approach to design work. The key design issues in the immediate future are the need for 'sustainable' design solutions and the impact of information technology, both in the creation of new products and new design tools.

Within this environment of continuous change, designers need to look forward and consider how they can improve their design process as part of improving the performance of the construction industry. Logically, before any work can start on improving the design process the current process needs to be defined. This will provide a number of benefits, including:

- establishing a common understanding of the design activities
- defining a minimum standard for good design practice
- clarification of the terms used in the various stages of the design
- improving communication; and most importantly
- confirming the starting point for improving the design process

The aim of the book is to provide the first step by defining the design process. Surprisingly the current building services design process does not appear to be explicitly defined in any existing texts. How the design process can be improved is a much more complex issue that will involve debate across the industry and some innovative thinking.

1.3 How to use the book

Although the book is written as a continuous work that can be read from start to finish, it is likely that practising designers who are short of time will use the book as a reference tool, dipping into the chapters infrequently as a type of check list. For these people, read Chapter 2 particularly the sections on responsibilities and then Chapter 3 as it provides an overview of the whole process. The other chapters can then be sampled as your work progresses through the various design stages. For those people who are studying the design process to understand how it works or to improve the process then the book will probably need to be read as a whole. Reference texts and further reading are provided at the end of each chapter and these will provide additional information, particularly the BSRIA documents.

Chapter 2

The building services designer's role and responsibilities

2.1 Chapter contents

The topics covered in this chapter include:

- An explanation of mechanical building services systems
- Designer's legal responsibilities
- Designer's contractual responsibilities
- Designer's professional responsibilities.

2.2 Mechanical building services systems

The range of building services systems a designer is likely to experience is directly reflected in the purpose of the employer's business. A company specialising in feasibility studies and concept designs for the retail market will have a very focused range of building services work, while a design and build company operating in all the market sectors will probably have a very varied range of work.

Different organisations also package different engineering subjects into mechanical building services engineering. For the purposes of this book the scope of building services is defined as:

- HVAC (heating, ventilating and air conditioning)
- Mechanical fire protection
- Utility services
- Noise control

These subjects are described in more detail below.

2.2.1 Heating, ventilation and air conditioning (HVAC)

Any system involved with heating, ventilating or air conditioning inside a building falls into this category [1]. The primary mechanical systems that can apply to any project are presented in Table 2.1.

Table 2.1 Typical HVAC systems

System	Description
Piped services	
Low temperature hot water (LTHW)	Heating plant and pipework distribution systems with radiators, fan coils or heater batteries, etc. These systems can be subdivided into either constant volume, variable temperature or variable volume, constant temperature
Domestic water services	Hot and cold water to basins, sinks, baths, showers, vending machines, etc. including potable and non-potable water systems
Chilled water	Chillers and pipework systems for distributing chilled water for cooling air before it is passed into an air conditioned or comfort cooled space and to cool process plant and equipment
Direct expansion refrigerant	A piped system transporting refrigerant acting like a freezer although operating in reverse to cool a space. Refrigerant systems extend from small individual room cassettes to cold rooms for product storage
Cooling tower water	Cooling towers or air blast coolers (indirect cooling coils) and distribution pipework frequently used to provide a heat sink for liquid cooled chiller plant
Steam and condensate	Steam generation and distribution pipework for heating, sterilising, and humidifying with liquid condensate recovery return pipework
Compressed air	Compressor equipment and piped compressed air for operating machines, control devices, etc.
Specialist water treatment systems	De-mineralised, reverse osmosis, distilled, de-ionised water, etc. and associated water treatment equipment
Specialist gas systems	Carbon dioxide, nitrogen, hydrogen, etc. Usually supplied through specialist subcontractors
Vacuum systems	Vacuum cleaning systems primarily used in industrial process and health care facilities. Usually supplied through specialist subcontractors
Above ground drainage	Soil waste drainage from sanitary accommodation
Above ground process drainage	Drainage from process plant, wash down areas and laboratories that will probably require on site treatment before discharge to the local authority sewage system
Rainwater drainage	Rainwater drainage from roofs discharging at ground level into below ground drainage systems
Natural gas	Natural gas pipework distribution from the service provider's connection
Ventilation services	
Mechanical ventilation	Supply and extract ventilation to occupied spaces
Toilet extract ventilation	Extract from toilets and change areas
Air conditioning	A ventilation system that controls the temperature and humidity in a space. Air conditioning systems have many different forms, dual duct, constant volume, etc.
Fume extract ventilation	Removal of contaminated air from processes generating fumes, e.g. work inside fume cabinets
Local extract ventilation	Removal of contaminated air from small point sources, e.g. solvent laden air from a chemical process, sampling point
Dust extract ventilation	Removal of dust from process plant handling solids, e.g. bagging of powders

Table 2.1 (continued)

System	Description
Natural ventilation	Ventilation of spaces utilising naturally occurring variations in pressure caused by temperature and wind
Kitchen extract	Extract from cooking equipment and kitchens
Clean area ventilation and pressurisation systems	Ventilation systems that ensure the particle count in a space does not exceed a set of defined criteria, e.g. pharmaceutical production areas. These systems often incorporate air conditioning and need to be designed in conjunction with the other particle shedding processes within the space
Car park extract	Extract from enclosed car parks
Hazardous area ventilation	(a) Pressurisation systems that segregate hazardous and non-hazardous areas, e.g. a manufacturing area handling flammable solvents and a process-control room. (b) Ventilation systems used to ventilate and control the space hazard classification
General	
HVAC controls	A system of instrumentation, control equipment and control devices, that automatically adjusts the plant and equipment settings to control the internal environment to within prescribed limits. Comprehensive packages can be extended into building management systems (BMS)
Noise control	HVAC noise control is not really a system; it is more of subsystem that needs consideration on all pipework and ventilation systems, e.g. attenuators, damping membranes, etc.

2.2.2 Mechanical fire protection

Fire protection engineering is generally split between the different engineering disciplines on the project (although some of the larger engineering and project management companies do combine the function within its own discipline, i.e. fire engineering). Passive fire protection, i.e. means of escape, fire compartmentation, is generally part of the architect's work. Fire detection, i.e. fire detection and alarm systems, is generally part of the electrical works and the mechanical fire protection systems usually become the responsibility of the mechanical building services designer.

The extent of the mechanical fire protection works are partly established through regulatory requirements and partly through discussion with the client, the clients' insurer. The outcome of the discussions are driven by the perception of the risk. Consequently the full extent of the fire protection requirements cannot be determined until these discussions have been held. However, the main mechanical fire protection systems that could apply on any project are presented in Table 2.2.

Mechanical fire protection engineering is generally outsourced to Loss Prevention Council (LPC) approved contractors (companies approved by the LPC to carry out fire protection design and installation work). The design requirements are then co-ordinated into the project work by the building services designer.

2.2.3 Utility services

The range of utility services that can be expected on any project are listed in Table 2.3.

Table 2.2 Mechanical fire protection systems

System	Description
Underground fire water mains	An underground fire water main served by a dedicated fire water pump set and water storage facility. The main supplies fire water to sprinkler, deluge, spill systems, etc.
Underground hydrant mains	An underground hydrant water main served by a dedicated hydrant water pump set and water storage facility. The main supplies water to private road level hydrant outlets for fire brigade use
Sprinkler protection	A charged piped system serving frangible sprinkler bulbs. In the event of a fire the bulbs subject to heat break and spray fire water over the surrounding area
Deluge protection	An open pipework system arranged around high risk plant items or areas, e.g. a vessel full of flammable solvent. A separate detection system activates the deluge valve. When the valve opens, fire water floods through the system and covers the whole area. These systems are often foam enhanced
Spill systems	A monitor or bleed system, foam enhanced and remotely activated, that seals a flammable liquid spill and prevents ignition
Smoke pressurisation[a]	A mechanical or natural ventilation system that, in the event of a fire, pressurises a fire fighting shaft (a staircase used by fire fighters) to prevent smoke contamination
Dry risers[a]	An open pipe system that is served from outside the building by the fire tender and runs up the fire fighting shaft. Fire fighters inside the building use the pipe to supply their hoses. Large buildings may have permanently charged 'wet' risers
Smoke extract[a]	A mechanical or natural ventilation system that, in the event of a fire, removes smoke from the contaminated zone
Hose reel[a]	A charged pipework system with hose reel outlets positioned around the building. In a fire condition they can be used to fight the fire
Hand-held fire extinguishers	Portable extinguishers used to fight a fire source before it becomes established. The type of extinguisher needs to match the likely fire type, e.g. paper, electrical equipment

a   These systems are usually procured through a HVAC work package although they are part of the mechanical fire protection works.

Many of these services are repeated from the HVAC section. The difference in application is defined by location. If the services are 'stand alone' inside an individual building they would be HVAC systems. If the services supply a building or a range of buildings they would be considered utility services. The water supply from the water company main in the street to the building would be a utility service and maybe installed before the building shell. The water supply inside the building to the kitchen sink would be a HVAC system and would be installed after the building was completed.

Utility services invariably involve a 'tie-in' into an existing system. Every tie-in installation requires special consideration as the details of the existing system are often unknown and tie-ins generally require system shut downs. Shut downs are

Table 2.3 Utility services systems

System	Description
Steam and condensate	The steam distribution pipework system from a new or existing site boiler house into the new facility
Potable water	The water supply from the water company main in the road to the new facility
Natural gas	The natural gas supply from the gas distribution company underground main to the new boiler plant or chillers
Chilled water	The chilled water flow and return pipework system from the clients remote chiller plant and into the building
Cooling tower water	Cooling tower water flow and return pipework system from the remote cooling tower plant and into the facility

problematic as the client or utility service provider generally does not want to lose production and they can reveal other system faults.

2.2.4 Workplace and environmental noise control

Health and safety, and environmental legislation determines the requirements for noise control engineering.

The Noise at Work Regulation 1989, enabled under the Health and Safety at Work Act 1974, limits the noise dose that a person can receive in any single day. The current legislation describes two action levels, i.e. noise dose levels that must promote an action by the employer. These limits are currently set at a first action level of 85 L_{eqA} and a second action level of 90 L_{eqA}. New plant design needs to ensure, with due regard to what is 'reasonably practicable', that the noise environment within the space does not exceed these limits.

The Environmental Protection Act (EPA) 1990 [2] includes noise that is 'prejudicial to health or a nuisance' as a statutory nuisance. To ensure that no statutory nuisance occurs the 'enforcing authority' (i.e. the environmental health officer or environmental inspector) commonly 'requests' that the new works do not contribute to the existing background noise level. This means that any new plant must not increase the surrounding background noise level beyond the level that existed before the plant was built. In practical terms the building services designer meets with the enforcing authority and agrees a number of noise reference points around the plant site boundary. The background noise level is the level recorded at these points before work starts on site. (The background noise level can generally be defined as the noise level exceeded 90% of the time, measured at the quietest time of day, usually between 02:00 and 04:00 h.) The design and installation of the new plant then takes place. During post construction commissioning, the noise levels are re-measured and provided the noise level has not increased the plant is acceptable. If the noise level has increased or the enforcing authority has received a noise related compliant (e.g. from a local resident) that can be attributed to the new plant then the plant can be shut down until the issue is resolved.

2.3 Building services design tools

Nowadays the design of the building services systems is supported by a range of tools. Typically these tools are:

- A telephone;
- A networked PC, providing access to a range of facilities such as:

  - word processing, spreadsheets and presentation software
  - steady state heat loss, heat gain software
  - pipework and ductwork sizing software
  - fluid flow modelling and thermal simulation software
  - access to the internet
  - CAD drawing software, usually AutoCAD
  - combined CAD drawing and calculation software;

- A technical catalogue library;
- A Barbour index CD ROM information system or equivalent, containing latest copies of most British Standards, Building Services Research and Information Association (BSRIA) information, Chartered Institution of Building Services Engineers (CIBSE) documents and relevant industry publications. These systems are fairly expensive and may not be available in all companies;
- As a minimum copies of the CIBSE design guides A, B and C and perhaps Institute of Plumbing (IOP) and American Society of Heating, Refrigeration and Air-Conditioning Engineers (ASHRAE) guides;
- Copies of previous project work including drawings and specifications;
- An installation specification defining installation quality standards for pipework, ductwork, insulation, testing and commissioning and operation and maintenance manuals (see Section 7.8.5);
- Perhaps most importantly the experience of other designers; very few technical problems are unique and it is likely somebody else has dealt with the issue previously.

2.4 Legal responsibilities

European and National health and safety legislation imposes specific legal duties on all designers.

Health and safety legislation is initiated either under Article 100A (products and equipment) or Article 118A (people and systems) of the Treaty of Rome and issued as European Union directives.

The directives set out objective standards that the member states have to achieve. The individual member states then decide how the requirements of the directive will be introduced into their own legal structures.

In the UK, the directives are implemented through two different routes, the European Communities Act 1972 (ECA) for Article 100A issues and the Health and Safety at Work Act 1974 (HASAWA) for Article 118A issues. Figure 2.1 provides an overview of the two routes.

	Headline safety legislation	
European Communities Act 1972		Health and Safety at Work Act 1974

	Typical Regulations issued under the headline Acts that provide a statement of the requirements	
• Supply of Machinery (Safety) Regulations 1992 • Boiler (Efficiency) Regulations 1993		• Pressure Systems and Transportable Gas Containers Regulations, 1989. • The Workplace (Health, Safety and Welfare) Regulations, 1992. • The Construction (Design and Management) Regulations, 1994. • The Control of Substances Hazardous to Health Regulations, 1994.

	Method of complying with the Regulatory intent.	
(1) Risk Assessments (2) Apply Hierarchy of Risk Control (3) Apply any specific Harmonised Standards e.g. BS EN 779 Particulate Air Filters		(1) Risk Assessments (2) Apply Hierarchy of Risk Control (3) Apply any specific Guidance Requirements e.g. HS(G) 39 Compressed Air Safety

	Legislation "target" group	
Article 100A - Products and Equipment		Article 118A - People and Systems

Note

a) Further detailed information on complying with the Health and Safety legislation in design is contained in Chapter 6 Concept Design.

b) Harmonised standards in the United Kingdom are referred to as BS EN documents

Figure 2.1 The application of health and safety legislation in the United Kingdom.

2.4.1 The European Communities Act 1972

In order to achieve the free movement of goods within the European Union, in 1985 the Council of Ministers adopted the 'New Approach to Technical Harmonisation and Standards' directive. This directive and subsequent similar directives are implemented through the ECA as they affect products and equipment.

The majority of the legislation implemented under the ECA aimed at harmonisation is specifically directed at equipment manufacturers and the use of CE marking. The CE mark is an abbreviation of 'Conformité Européen', i.e. European Conformity. When the manufacturer attaches a CE mark to a piece of equipment or product they are claiming that it is safe and that it complies with all the appropriate legislation.

The main legislation issued through the ECA that affects building services designers work is the Supply of Machinery (Safety) Regulations 1992. The regulations apply to all new machinery manufactured or supplied in the UK and intended for use in the European Economic Area (e.g. all the EU and EFTA countries except Switzerland). The term 'machinery' is very broadly defined. In this context machinery includes the design and installation of mechanical building services systems. Each system is a unique mix of CE marked plant and equipment, e.g. fans, pumps, joined together with pipes, ducts and cables forming a new independent machine. The regulations [3]

> require all UK manufacturers and suppliers of new machinery to make sure that the machinery they supply is safe. They also require manufacturers to make sure that:
>
> - machinery meets relevant essential health and safety requirements (these are listed in detail in the regulations), which include the provision of sufficient instructions;
> - a technical file for the machinery has been drawn up, and in certain cases, the machinery has been type-examined by a notified body;
> - there is a 'declaration of conformity' (or in some cases a 'declaration of incorporation' for the machinery, which should be issued with it...;
> - there is a CE marking affixed to the machinery (unless it comes with a declaration of incorporation).

The range of penalties for contravening the regulations consist of a fine of up to £5,000 or imprisonment for up to 3 months.

2.4.2 The Health and Safety at Work Act 1974

The Act allows the Secretary of State to create a system of regulations and approved codes of practice that progressively replaces existing legislation and maintains or improves the standard of health, safety and welfare (some pre-1974 prescriptive legislation, e.g. Factories Act 1961 is still in place however most of its content has now been repealed and replaced with new legislation under the HASAWA). The Act is an 'enabling' act, a broad article of legislation that does not go into any great detail. The detail is provided in the regulations and approved codes of practice enabled under the Act, as these can more readily be adapted to keep pace with developments.

The range of the regulations and approved codes of practice enabled under the HASAWA is huge. Perhaps the most frequently occurring regulations that impact on building services design are:

- The Electricity at Work Regulations, 1989
- The Noise at Work Regulations, 1989
- Pressure Systems and Transportable Gas Containers Regulations, 1989
- The Workplace (Health, Safety and Welfare) Regulations, 1992
- The Construction (Design and Management) Regulations, 1994
- The Control of Substances Hazardous to Health Regulations, 1994
- The Provision and Use of Work Equipment Regulations II 1998

Invariably the approved codes of practice provide more objective guidance on complying with the regulatory intent. This guidance and a raft of other excellent health and safety documentation is published by the Health and Safety Commission (HSC). (The HSE have a very useful 'keyword' search engine on their home page at http://www.open.gov.uk/hse/hsehome.htm.)

The HASAWA [4] imposes a number of duties on designers. Section 6 of the act places a duty on

> Any person who designs, manufactures, imports or supplies any article for use at work…to ensure, so far as is reasonably practicable, that the article is so designed and constructed that it will be safe and without risks to health…that persons supplied by that person are provided with adequate information about the use for which the article is designed…and about any conditions necessary to ensure that it will be safe and without risks to health.

Contravention of the duties imposed under the Act as a designer is likely to result in a fine of up to £5,000 or imprisonment of up to 6 months. However penalties as an employer and for companies can be significantly more severe.

2.4.3 The Construction (Design and Management) Regulations 1994

The Construction (Design and Management) Regulations (CDM) are enabled through the HASAWA and automatically carry the designers' duties described in the section above. Unlike most of the other regulations enabled under the act they are not specific to a single technical issue. They apply to almost all construction design activities and impose specific duties on a wide range of people involved in the design and construction process.

The HSC [5] guidance on the regulations specifically defines 'designers' to include building services designers and states that the 'designer' may be an individual, partnership or firm employing designers. The designers duties under the regulations are to:

- Make the client aware of their duties
- Identify the significant health and safety hazards and risks of any design work
- Give adequate regard to the hierarchy of risk control (see Fig. 5.3)
- Provide adequate information on health and safety to those who need it
- Co-operate with the planning supervisor and, where appropriate, other designers involved in the project

Anyone engaging a designer needs to ensure that the designers are aware of their duties and have the competence and resources to comply with them.

2.4.4 Further health and safety information

Further information on satisfying the designers' legal duties defined in The Supply of Machinery (Safety) Regulations 1992, The Health and Safety at Work Act 1974 and The Construction (Design and Management) Regulations 1994 is provided in Chapter 5.

2.5 Contractual responsibilities

An organisation's contractual responsibilities on any project are whatever the organisation accepted when it took on the contract! Although this is obvious it is also very informative as the responsibilities on each project can vary significantly; the requirements are only those detailed in the contract and if the contract details are not clear and precise, conflicts are likely to occur during the execution of the project work.

To reduce confusion the professional bodies in the construction industry publish their own standard agreements. The building services industry tends to use the Association of Consulting Engineers (ACE) conditions of engagement [6] as a basis for the provision of services.

The ACE publish different conditions of engagement to suit different applications:

Agreement A: for use where a consulting engineer is engaged as a lead consultant
Agreement B: for use where a consulting engineer is engaged directly by the client, but not as lead consultant
Agreement C: for use where a consulting engineering is engaged to provide design services for a design and construct contractor
Agreement D: for use where a consulting engineer is engaged to provide report and advisory services
Agreement E: for use where a consulting engineer is engaged as a project manager
Agreement F: for use where a consulting engineer is engaged to act as planning supervisor in accordance with the Construction (Design and Management) Regulations 1994

ACE conditions of engagement, agreement B (2), for mechanical and electrical services in buildings has three main sections

A: Memorandum of agreement
B: Conditions of engagement
C: Schedule of services

Perhaps the most important aspects of the document from the designers view point are the memorandum of agreement and the schedule of services.

The memorandum of agreement in clause A7 describes the scope of work. This is a general list of systems, e.g. air conditioning and mechanical ventilation services, that could form part of any project. The list should be edited to suit the specific project requirements by deleting items and adding others that are unique to the new work.

The schedule of services is broken down into two main sections, normal services and additional services. The normal services schedule lists the designer's services and places them in nine subsections. Each subsection represents a step in the design development process and is intended to harmonise with the Royal Institute of British Architects' (RIBA) plan of work (a plan of work that breaks down the design and construction process on a generic project into different stages, see Fig. 3.1, and Section 3.2.10 Links with the ACE normal services stages). The additional services section lists services that are considered extra to normal project requirements. Where specific project requirements need some or all of the additional services, they are simply retained in the agreement otherwise they are deleted.

The other conditions of engagement are similarly tailored to specific project requirements by adding and omitting elements in the scope of work and schedule of services.

There are many other forms of contract between consultants and clients such as RIBA standard form of agreement (primarily for architectural commissions) and Federation Internationale des Ingenieurs-Conseils (FIDIC) conditions of contract. On very small works it may only be a letter between the parties explaining the scope and the fee.

Design and build projects operate different forms of contract as the scope of work requirements include installation. Typically these types of projects use the IChemE Red Book or Joint Contracts Tribunal (JCT) 1980 standard form with contractors design.

The essential issue, irrespective of the form of contract, is to understand the contractual requirements on the project and any internal company split in the scope of work responsibilities between different engineering disciplines, e.g. fire, electrical, etc.

2.5.1 Construction project contracts

The extent of any designers contractual duties will be heavily influenced by the client's contract strategy for the project. Although there are many derivatives of procurement options for clients, the four main approaches are: the traditional contract, design and build, management contracting and construction management.

2.5.1.1 The traditional contract

The project is designed and managed by the design team (architect, structural engineer, building services designer, etc. to suit the project requirements) and constructed by a main contractor, with a range of subcontractors (Figure 2.2).

Each member of the design team, the main contractor and primary subcontractors,

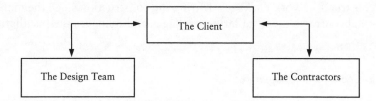

Figure 2.2 The contractual relationship on a traditional contract.

e.g. the building services contractor, will have their own contract with the client. The contractors tendering process can involve a single stage or two stages. A single stage tender is based on a complete design and the contractor submits a single bid for the work. Two-stage tendering is preferred when there is an obvious benefit in integrating the contractors knowledge into the design before it is complete, e.g. short design period, unique construction approach. Two-stage tendering involves the contractor in costing the work against the incomplete design and submitting a revised cost when the design is complete.

2.5.1.2 Design and build

The contractor is responsible for designing and building the project. The design team and any subcontractors are integrated into the contractors project organisation (Figure 2.3).

Prior to the client establishing a design and build contract, it is very likely that a design team will have prepared a technical performance specification as a basis for contractor tendering. In these circumstances it is possible that the client may novate the original design team into the contractors team or retain them to oversee the contractors work.

Other derivatives on the design and build theme include:

- turnkey contracts – where the contractor constructs the project and fits it out with all the necessary equipment, e.g. tables, chairs, computers, etc.
- private finance initiative (PFI) – similar to turnkey contracts though the contractor manages and operates the facility for a defined concession period

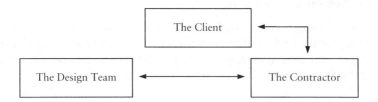

Figure 2.3 The contractual relationship on a design and build contract.

2.5.1.3 Management contracting

The design team is extended to include a management contractor. The management contractor injects additional construction expertise into the design process and organises the construction work packages. During the construction stage the management contractor subcontracts the installation packages and manages the installation work (Figure 2.4).

2.5.1.4 Construction management

The construction management contract is similar to the traditional contract route. The design team is extended to include a construction manager who provides up-front

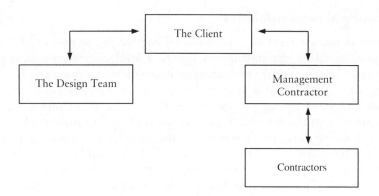

Figure 2.4 The contractual relationship on a management contract.

construction advice and guidance on the construction packages. During the construction stage they manage the site. Contractors involved in the installation work are contracted directly to the client (Figure 2.5).

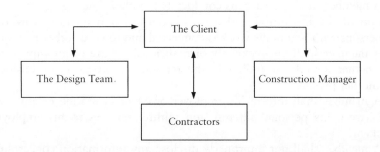

Figure 2.5 The contractual relationship for construction management.

2.5.2 *The effect on the designer*

Although there are numerous routes for clients to procure construction projects, the building services design process essentially remains the same irrespective of the route. The design process starts with the client recognising a need for a new facility and ends when the facility is fully functional. The variation in the building services designers' role is where they start in the process and where their involvement ends. A consultant's role on a traditional contract will require a continuous involvement from the outset of the project through to project completion. On a design and build contract a consultant's role may be confined to front end work and development of performance deliverables (e.g. drawings, specifications, etc.). A designer's work for a contractor on a construction management contract would involve working up the construction design information and resolving plant and equipment details.

2.6 Professional responsibilities

Most designers are members of a professional engineering institution. Professional qualifications issued by the institutions signify a level of academic and practical achievement and a commitment to continuing professional development. Registration on the engineering council as a chartered engineer, an incorporated engineer or an engineering technician is only possible for designers who are members of a professional institution that is a nominated body of the council. The main nominated institutions for building services designers are the Chartered Institution of Building Services Engineers (CIBSE) and the Institution of Mechanical Engineers (IMechE).

Both institutions require their membership to demonstrate professional conduct in their duties. CIBSE [7], Professional Conduct by-law 38 states: 'A member shall at all times so order his conduct as to uphold the dignity and reputation of his profession, and to safeguard the public interest in matters relevant to the science and practice of building services engineering. He shall exercise his professional skill and judgement to the best of his ability and discharge his professional responsibilities with integrity.'

The requirements of the IMechE [8] are described in by-laws 33 and 34. By-law 33 code of conduct for all members states:

33.1 A member shall so order his conduct as to uphold the dignity and reputation of the Institution and act with fairness and integrity towards all persons with whom his work is connected and towards other members.

33.2 A member shall not recklessly or maliciously injure or attempt to injure whether directly or indirectly the reputation, prospects or business of another person.

33.3 A member shall inform his employer or client in writing of any conflict between his personal interest and faithful services to his employer or client.

33.4 A member shall not improperly disclose any information concerning the business of his employer or client or of any past employer or client.

33.5 A member shall not solicit nor shall he receive any inducement directly or indirectly on any article or process used in or for the purpose of the work in respect of which he is employed unless or until such inducement has been authorised in writing by his employer or client. Gifts of a relatively trivial nature are not considered to be an inducement.

33.6 A member shall at all times so order his conduct as to safeguard the public interest, particularly in matters of health and safety and the environment.

33.7 A member whose advice, based on his skill and judgement, is not accepted shall take all reasonable steps to ensure that the person overruling or neglecting his advice is aware of any danger which the member believes may result from such overruling or neglect.

33.8 A member who is a student, graduate, associate member or member, but not a member who is substantially retired, shall strive to progress through the classes of Institution membership.

By-law 34 additional code of conduct for corporate members states:

34.1 A corporate member shall so order his conduct as to uphold the dignity and reputation of his profession.

34.2 A corporate member shall not solicit nor accept a consideration in connection with professional services rendered to his employer or client other than from his employer or client or with his employer's or client's consent.

34.3 Continuing professional development (CPD). Every corporate member who practices the profession of mechanical engineering is required to maintain and develop relevant professional competence to the standard prescribed in the qualifications and training regulations established by the council pursuant to by-law 17.1.

Both institutions have disciplinary procedures for improper conduct by a member. Improper conduct extends beyond a simple breach of the by-laws and includes broader issues such as bringing the institution into disrepute. The IMechE list a number of specific activities that constitute improper conduct; these include any breach in the by laws, gross negligence, gross technical incompetence, recklessness in a professional respect and conviction by a competent tribunal of any criminal offence. Penalties include expulsion from the institution and loss of engineering council registration.

2.7 Chapter review

This chapter explains the designer's role and responsibilities through the following subjects:

1 Mechanical building services systems:
   ● building services systems are divided into four sections: HVAC, mechanical fire protection, utilities and noise control
   ● each section outlines the typical systems that may involve design work by building services engineers;
2 Engineer's legal responsibilities:
   ● health and safety design responsibilities as described in current European and UK legislation
   ● the specific design requirements of the Supply of Machinery (Safety) Regulations 1992, the Health and Safety at Work Act 1974 and the Construction (Design and Management) Regulations 1994;
3 Engineer's contractual responsibilities:
   ● the ACE form of agreement and application
   ● the main construction project contract forms: traditional contracts, design and build, management contracting and construction management;
4 Engineer's professional responsibilities:
   ● CIBSE and IMechE codes of conduct.

References

[1] *A Design Briefing Manual*, Building Services Research and Information Association Publications, 1990, Application Guide 1/90.

[2] *Pollution Handbook*, National Society for Clean Air and Environmental Protection Publications, 1995.
[3] *Supplying New Machinery* INDG270, HSE Books.
[4] *Health and Safety at Work Act*, HMSO.
[5] *A Guide to Managing Health and Safety in Construction*, HSE Books.
[6] *Conditions of Engagement*, Second Edition, The Association of Consulting Engineers Publications, 1998, Agreement B (2).
[7] *Royal Charter and By-Laws*, The Chartered Institution of Building Services Engineers, 1996.
[8] *Code of Conduct*, Institution of Mechanical Engineers.

Further reading

Holt, A.J., *Principles of Health and Safety at Work*, The Institution of Occupational Safety and Health Publishing, 1995.
Architects Job Book, Sixth Edition, RIBA Publications, 1995.
Allocation of Design Responsibilities for Building Engineering Services, Building Services Research and Information Association Publications, Technical Note TN 21/97.
Project Management Handbook for Building Services, Building Services Research and Information Association Publications, Application Guide AG 11/98.
Designing for Health and Safety in Construction, HSE Books.

Chapter 3

Design in engineering construction

3.1 Chapter contents

The topics covered in this chapter include:
- An overview of the design process (detailed information on the individual stages in the design process is provided in future chapters)
- Design responsibility

3.2 The design process

The design process on all projects follows the same stages of development. However the extent and detail of the activities behind each stage are different with every project.

There are a number of different industry formats for mapping the stages in the engineering construction process [1,2]. A sequence that provides a simple fit with building services design activities is outlined in Figure 3.1. This is based on the RIBA Plan of Work [3] and the normal services stages used in the ACE Conditions of Engagement [4]. The stages highlighted in bold represent activities of building services designers.

The design process has seven stages: feasibility, concept design, scheme design, detail design, construction design information, construction and feedback; while the briefing process has a single stage, the design brief. Earlier stages in the briefing process involving initial client discussions and definition of the brief are generally led by the project manager (architect, quantity surveyor, etc.) or prepared by the client. These activities do not generally involve significant design input though they frequently benefit from building services technical advice (see Chapter 5).

An overview of each stage in the design process together with its typical building services design activities and deliverables is provided below.

3.2.1 Feasibility studies

Feasibility studies evaluate specific technical issues on project proposals to determine if they could prevent the project from continuing. Proposals fail because they are not technically possible or they are technically possible but the cost or quality or time scale are not acceptable. Invariably feasibility work rather than stopping the development of a project involves evaluating numerous project options to determine the optimum solution.

If the project proposal is easily defined then a feasibility study may not be required.

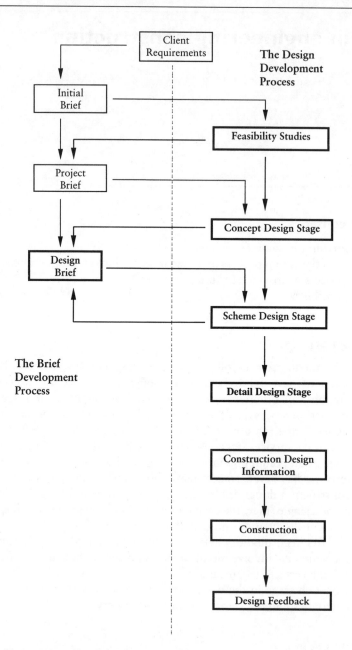

Figure 3.1 The process of brief and design development for building services.

In this instance the design process would start at the concept design stage.

Typical feasibility study activities

- Review the initial brief material
- Survey the proposed site and determine any site constraints, e.g. space restrictions

for distribution of services, limitations on noise emissions, site contamination, etc.

- Contribute to project team meetings and brain storming
- Advise the client of their CDM duties
- Establish the clients key success criteria for the project
- Identify any significant health and safety hazards and design work risks
- Evaluate the technical design impact of the different project options, e.g. utility services demand, future expansion, etc.
- Evaluate any specific technical issues that may prevent the project from proceeding
- Provide input to the project evaluation and value management process
- Contribute to the project programme
- Prepare the feasibility study report

Feasibility study design deliverables

- Feasibility study report

3.2.2 Concept design stage

Concept work uses the project brief as a basis to develop a number of suitable concept design options. This involves developing the design information, determining the influence of client's key success criteria on the design and ensuring the technical solutions comply with statutory requirements. Finally, all the concept design options are evaluated to determine the best solution.

Typical concept design activities

- Review the project brief and feasibility study information
- Review similar previous projects and feedback information
- Survey the proposed site and determine any site constraints (if this was not carried out in the previous stage)
- Contribute to project meetings, design team meetings and brain storming sessions
- Establish the internal and external environmental design criteria, the life span of the facility and any 'essential' system requirements, e.g. cooling systems for electronic equipment rooms, future expansion requirements, etc.
- Generate and develop a number of concept design options
- Identify health and safety hazards, carry out risk assessments and where necessary apply the hierarchy of risk control (CDM regulations requirements)
- Contribute to the project programme
- Interpret the affects of the proposed building fabric and where appropriate the impact of the enclosed process on the concept design, e.g. lobbies between hazardous and non-hazardous areas, the requirements for HVAC plant rooms, services distribution riser shafts, fire fighting shafts, etc.
- Discuss issues related to the concept design options with the appropriate authorities, e.g. fire officer, environmental health officer, etc.
- Prepare a preliminary ±30 per cent cost estimate for the concept design options on

a cost per m² basis or by elemental cost break down
- Evaluate the concept design options and determine the best solution
- Determine the likely impact on the project of statutory legislation and any specific client technical requirements, e.g. health and safety legislation, building regulations, insurance requirements, fire protection solutions, client standards, etc.
- Co-ordinate and communicate with the other members of the design team, e.g. advise the architect on the need and size of the ceiling voids, etc.
- Advise on the risks associated with the concept design
- Prepare appropriate concept design drawings
- Prepare the concept design report

Concept design deliverables

- Concept design report
- A ±30 per cent cost estimate
- Concept design drawings

The concept design stage is complete when the client signs off the concept design deliverables.

3.2.3 The design brief document

The design brief is the final stage in the briefing development process. The briefing process has refined the original, high level, client requirements into a definitive discipline-specific, statement of the work requirements.

Typically the building services design brief details the following information:
- A description of the project
- The clients success criteria for the project – specific to building services
- A summary of the building services scope of services provided by the designers
- A summary of the project's building services scope of work
- The basis for design development, i.e. the project design standards and design criteria
- A list of the design deliverables
- Confirmation of any other project specific criteria, e.g. energy efficiency issues, environmental issues, design margins, etc.
- The design acceptance criteria
- An outline programme for the design work

If a project is going to be successful, the design brief must be agreed and signed off by the client and the project team before the start of scheme design.

It is likely that further design development during scheme design will require minor revisions to the design brief. Under these circumstances the document must be updated, reissued and signed off.

3.2.4 Scheme design stage

Scheme design develops the concept design information into a workable solution. This is achieved by developing and detailing the concept design and absorbing the specific requirements of all parties involved in the project. It is a very iterative process involving design, review, discussion and revision until the design solution is substantially complete. Where appropriate the scheme design is value engineered to improve compatibility with the clients key success criteria.

Typical scheme design activities

- Review the concept stage deliverables and design brief
- Contribute to project and design team meetings
- Communicate and co-ordinate the design with the other members of the project team, e.g. confirm plant weights, estimate electrical loads, etc.
- Integrate the other design team members' work into the design solution, e.g. revise the layouts to suit the updated structural details, etc.
- Develop preliminary system schematics
- Carry out calculations to allow primary plant selections
- Generate and develop scheme layout drawings
- Establish the primary building services distribution routes and plant space requirements
- Identify health and safety hazards, carry out risk assessments and where necessary apply the hierarchy of risk control (CDM regulation requirements)
- Contribute to the project programme
- Review the design with all external third parties, e.g. building control, the health and safety executive, the environmental health officer, public utilities etc.
- Determine the precise requirements of the applicable statutory legislation and client standards, e.g. sprinkler protection, minimum ventilation rates, health and safety risk assessment outcomes, material finishes, etc.
- Discuss with the primary plant and equipment suppliers the individual requirements of their plant, e.g. chiller and boiler suppliers, to establish any unique restraints
- Develop the scheme design with all appropriate specialist subcontractors, e.g. controls specialists, specialist fire protection contractors, etc.
- Value engineer the scheme design
- Provide a ±15 per cent cost estimate for the scheme design solution using supplier quotations for plant and equipment and contractor quotations for the installation works (pipe and duct work) or an estimate based on priced schedules of rates from previous projects
- Review the concept design technical risk issues and if possible design them out, e.g. increase plant maintenance access space, design the system for the short construction programmes (e.g. packaged and skid-mounted plant, reduced ductwork), thermally model the system proposals, etc.
- Carry out technical risk assessments on any 'essential' systems, e.g. air conditioning systems that influence product manufacturing, chiller water systems controlling computer suites, etc.
- Update the design brief document to reflect any design changes
- Determine the content of the building services work packages, e.g. the number of specialist subcontracts, pre-purchase of long delivery items, segregation of ductwork and pipework contracts, etc.

- Prepare the performance and installation specifications or a scheme design report (see below)
- Prepare a tender list of contractors

Scheme design deliverables

- Scheme design report (all contracts except design and build and specialist subcontractor works)
- Performance and installation specifications (design and build contracts and specialist subcontractors work only)
- A ±15 per cent cost estimate of the scheme design
- Scheme design drawings
- A tender list of contractors (design and build contracts only)
- Update and re-issue the design brief (as necessary)

Design and build contracts and specialist subcontractor works (e.g. fire protection) require a performance specification as a scheme design deliverable. When detail design forms part of the designer's work only a scheme design report is required.

The scheme design stage is complete when the client and project team agree the design and sign off the scheme deliverables.

3.2.5 Detail design stage

Detail design involves providing further detail to the content of the scheme design deliverables. The final design development work should have been completed during scheme design.

Typical Detail design activities

- Review the scheme design deliverables
- Ensure that the detail design is compatible with the agreed design acceptance criteria
- Contribute to project and design team meetings
- Communicate and co-ordinate with the other members of the project team, e.g. prepare builders work information, confirm the final electrical loadings, etc.
- Integrate the other design team members work into the design solution
- Develop the detail design schematic drawings
- Carry out detail design calculations
- Select all the plant and equipment associated with the building services systems
- Develop the detail design general arrangement layout drawings
- Generate and develop the detail design plantroom plan and elevation drawings
- Generate and develop any design standard details that apply across the building services project work
- Discuss with the suppliers the individual requirements of their plant and equipment and how these requirements can be integrated into the project work
- Establish all the building services distribution routes and plant space requirements.

- Integrate any specialist subcontractors' packages into the project works
- Identify health and safety hazards, carry out risk assessments and where necessary apply the hierarchy of risk control (CDM regulation requirements)
- Contribute to the project programme
- Review any remaining design risks and where possible design them out, i.e. design margins, remote (BMS) monitoring during the post-commissioning period, contractors' design responsibilities, etc.
- Prepare the building services detail design and installation specifications
- Prepare a tender list of suitable contractors
- Prepare the plant and equipment schedules
- Review and confirm a ± 10 per cent cost estimate for the detail design solution

Detail design deliverables

- Detail design and installation specifications
- Plant and equipment schedules
- Detail design intent drawings
- Confirm a ± 10 per cent cost estimate for the detail design
- A tender list of contractors
- Update and re-issue the design brief (as necessary)

The detail design stage is complete when the client and project team agree the design and sign off the detail design deliverables. Generally all the non-technical aspects of the work packages, e.g. the form of contract, invitation to bid letters, preliminaries, etc. are provided by the quantity surveyor. The combined pack of information forms the contract documentation and is issued for tender to the contractors on the tender list. Generally, after the returned tenders are analysed the contractors that submit the best tenders are invited for a tender interview. Subject to client agreement the design process then moves to the next stage and the successful contractor is awarded the contract.

3.2.6 Construction design information

During the construction design information stage, the designer's role changes from design development and production to design verification and support. The designer assists the contractor in developing the design. The contractor produces the construction design information, detailing how the design will fit into the space and how it will be fabricated.

Typical construction design information activities

- Review the detail design deliverables
- Review of the contractors' construction design information; this typically consists of:
  - ductwork fabrication drawings
  - pipework layout drawings
  - plantroom layout drawings
  - plant and equipment manufacture's detail information

      – control and instrumentation drawings
      – builder's work construction detailing
      – a construction programme

- Answer the contractors design queries
- Confirm the extent of satisfactory completed design work for valuation
- Monitor the contractors' progress against the programme
- Review the contractors' installation method statements
- Contribute to project and site team meetings
- Review and confirm factory acceptance testing of equipment, e.g. air handling unit off-site testing, fume cabinet 'mock-ups' tests, etc.
- Communicate and co-ordinate any design changes with the other members of the project team

The extent of the activities at this stage are determined by the project contract and how the work packages have been let. When the work has been let as an installation contract the construction design information activities are typically those listed above.

When the work has been packaged as a design and build or specialist performance subcontract, the detail design work will be carried out by the contractor or the specialist subcontractor. In these circumstances the designer who specified the original works will, to the best of his/her ability, need to ensure that the design is correct. To help ensure that the design is correct (in addition to choosing a suitable contractor) it is beneficial to specify in the contract documentation, the supply of a range of design deliverables. These design deliverables are project-specific, typically consisting of calculations and design information used in the contractor's design work.

3.2.7 Construction stage

Design activities in the construction and commissioning stage are essentially policing and witnessing. The designer reviews the contractors' installation works to ensure it meets the performance and quality requirements described in the detail design deliverables and construction design information. On completion of the installation, the systems are commissioned and set to work to prove the design and confirm the design acceptance criteria.

Typical construction activities

- Review the contractors' construction design information and design deliverables
- Review the contractors' quality plan
- Monitor the contractors' progress against the programme
- Agree the installation work that is included in interim valuations
- Ensure the installation quality is acceptable
- Witness and sign off site testing of the systems
- Answer the contractors' site queries
- Review the contractors' installation method statements
- Contribute to project and site team meetings

- Ensure the commissioning technical information is complete
- Witness and sign off the system commissioning
- Ensure the installation meets the design acceptance criteria
- Review and agree the operation and maintenance information

3.2.8 Feedback stage

Feedback is a critical part of the overall design process. It reinforces the good aspects of the design solution and highlights areas for improvement. The process is a structured method for collecting and integrating feedback into future design solutions.

Feedback activities

- Seek feedback information from the client, project team and contractor
- Carry out an internal design team feedback workshop
- Record and sort feedback information for future use

3.2.9 Presentation of the design process

Mapping the design process gives the impression that the design development occurs in well-defined stages, with one finishing before the next starts. This can be a little bit misleading. All design work follows this process although in many cases the definition between stages can be blurred. It is essential that the early stages in the design process are clearly defined and closed out before proceeding to the next stage. However, frequently the production of the construction design information and the construction stage overlap as the need to start on site to meet the overall programme becomes critical. This is quite acceptable provided the work has been broken up into sections and the construction design information for any single section is reviewed and agreed before construction starts. If the construction activities start without the agreed construction design information then the project proceeds at risk. If it transpires that the installation is incorrect the contractor may be entitled to claim a programme delay and associated costs. This frequently leads to a deterioration in the relationship between all the parties as programme delays and increased costs jeopardise current and future project success. Consequently clarification and agreement of the construction design information prior to construction is essential.

3.2.10 Links with the ACE normal services stages

The stages in the brief and design development process can be cross-referenced with the ACE normal services stages. The relationship between the two models is described Table 3.1.

3.3 Design responsibility

Design responsibility in the engineering and construction industry is invariably linked with design failure. Establishing who is responsible for a design failure is determined

Table 3.1 The brief and design development process and ACE normal services

Brief and design development stages	ACE normal services
Feasibility studies	Strategic briefing stage
Concept design stage	Outline proposals stage
Design brief	Part of the outline proposals stage
Scheme design stage	Detailed proposals stage
Detail design stage	Final proposals and production information stages
Construction design information	Part of the mobilisation, construction and completion stage
Construction	Part of the mobilisation, construction and completion stage
Feedback	No similar provision in normal services stage

by establishing who has design responsibility and whether they failed to provide the appropriate standard of design.

Design responsibilities should be clearly defined in the conditions of engagement between designer and the client. Where these agreements are not clear the courts appear to start with the viewpoint that the designers are entirely responsible for the design [5].

The ACE conditions of engagement state the extent of the consultants' design responsibilities. The consultant is only responsible for his own design work and that of any specialist subconsultants employed by them. Design work by other consultants, contractors and subcontractors where they have been employed with the clients' specific consent is excluded. If any part of the design work is let to another party without the clients' consent then responsibility for the design will remain with the consultant (Moresk Cleaners, Ltd. versus Hicks 1966). However, even if the clients' consent has been given there remains an obligation on the consultant to ensure that there are no defects in the design that are sufficiently obvious that the consultant should have spotted them (Investors in Industry versus South Bedfordshire DC 1986) [5].

The other aspect of design failure is the standard of design. English law expects providers of goods and services to satisfy one of two mutually exclusive standards, either:

- That the goods or services be fit for their purpose or
- That the provider exercises reasonable standards of professional care and skill in carrying out the services

One or the other standard applies as the tests for compliance are different. A fitness for purpose obligation is strict in law, i.e. if the goods or services are not fit for purpose then the provider is liable irrespective of whether or not they exercised reasonable care and skill. The reasonable care and skill test is fault based, i.e. the provider must be negligent in performing his duties to be liable, consequently this is less onerous.

The law imposes a reasonable care and skill test on all professionals. This require-ment is explicitly stated in the ACE conditions of engagement. 'The Consulting Engi-neer shall exercise reasonable skill, care and diligence in the performance of the Services' (B2, Clause 2.3). This approach places the emphasis on the client to establish the consultant's negligence. However, the law tends to treat design and build contractors differently. The provider of an actual building is viewed as similar to a seller of goods (Viking Grain Storage, Ltd. versus T.H. White Installations, Ltd. 1985). In these circumstances a fitness for purpose standard would be applied.

3.3.1 Concept and scheme design responsibility

The design deliverables are the outcome of all the information exchanged between the client, the project team and the designer.

In the unlikely event that the client has agreed and signed off design deliverables that define a new chocolate packaging plant and the client really requires a new retail park development then the design is right – the client's brief was incorrect!

If the client's brief asked for a new pharmaceutical plant that requires a manned operation Class K (BS 5295:1989) clean room and the design clearly does not meet this requirement then the designer would be liable for the design failure. (Unless during the design process they addressed the issue and agreed with all parties that a Class K standard is not a requirement.)

When the project works only extend to concept and scheme design (i.e. design and build contracts), future detail design is the responsibility of the design and build contractor carrying out those works.

3.3.2 Detail design responsibility

For consultants, the opportunity for error in the detail design stage is significantly higher than concept or scheme design. This is inevitable as the amount of detail provided is, by its nature, substantially larger than in the other design stages. They need to correctly specify all the pipe sizes, the duct sizes, the plant, the equipment, the material requirements and how all the components fit together in the space, etc. If any of the detail design is incorrectly specified or the system layout detailed on the drawings cannot work then it is likely the contractor will be able to claim a technical variation. The extent of the error will determine whether the consultant has been negligent. (Most larger contracts will incur some degree of site technical variations. A technical risk budget of 5 per cent of the contract value could be considered a reasonable maximum for any unforeseen site issues.)

The consultant's risks can be mitigated through providing the right level of detail design, communicating it clearly and establishing an internal design auditing process.

Design and build contractors have more flexibility in the detail design stage as they act as their own consultant and contractor. Consequently errors in their design highlighted by their own auditing process or the clients' consultants comments should be easier to accommodate.

3.3.3 Construction design information responsibility

The contractor is responsible for producing the construction design information. However, the terms of the contract determine the extent of the contractors' design responsibility for the content of the information. If the consultant has carried out the normal services duties (as described in the ACE conditions of engagement) then the contractors' design responsibilities will extend to supports, fixings and finalising the location of the services within the space confines established by the consultant. Where the consultant has performed additional services that include co-ordination drawings then the contractor's design responsibilities should only extend to supports and fixings. The key to successfully establishing the contractors' design responsibilities is a clear definition of responsibilities in the contract documents.

Design responsibility on design and build contracts remains with the contractor. In these circumstances the project contract strategy determines the extent of the construction design information review. Where a consultant has produced a performance specification, they may be required to confirm that the design is in accordance with the specification intent. If the design and build contract is let by a management contractor, they may want to extend the review to ensure that the design solution is robust and that the construction requirements are consistent with the project restraints.

3.4 Chapter review

This chapter describes the building services design process in engineering projects and design responsibility through the following subject areas:

1   The main stages in the design and brief process:
    - Feasibility
    - Concept design
    - The design brief
    - Scheme design
    - Detail design
    - Construction design information
    - Construction
    - Feedback
2   Design responsibility:
    - Concept and scheme design
    - Detail design
    - Construction design information

The chapter provides the basis for more detailed discussion on the design and brief process in future chapters.

References

[1] *Design Management Systems*, BSI Publications, 1994, BS 7000: Parts 3 and Part 4.
[2] *Generic Design and Construction Process Protocol*, EPSRC, IMI.
[3] *Architects Job Book*, Sixth Edition, RIBA Publications, 1995.
[4] *Conditions of Engagement*, Second Edition, The Association of Consulting Engineers Publications, 1998, Agreement B (2).
[5] *Allocation of Design Responsibilities for Building Engineering Services*, Building Services Research and Information Association Publications, Technical Note TN 21/97.

Further reading

Feedback for Better Building Services Design, Building Services Research and Information Association Publications, Application Guide AG 21/98.

Project Management Handbook for Building Services, Building Services Research and Information Association Publications, Application Guide AG 11/98.

Design Information Flow, Building Services Research and Information Association Publications, Technical Note TN 92/17.

Further reading

Chapter 4

Feasibility studies

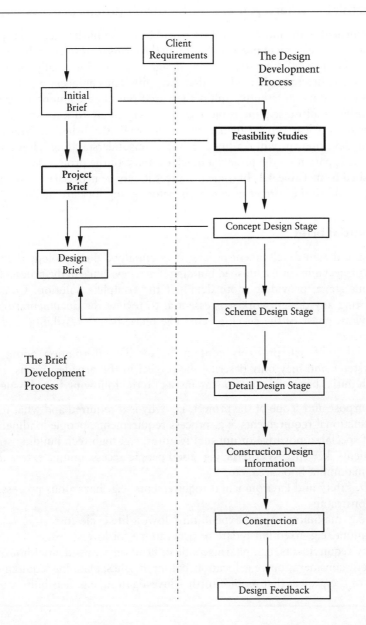

4.1 Chapter contents

The topics covered in this chapter include:

- An explanation of initial brief information
- Feasibility studies
- Project evaluation and value management
- An explanation of project brief information

4.2 Determining where you are in the development process

In practice probably the most difficult issue will be to establish where your particular project is in the development process. Although the terms used to describe the stages in the process, i.e. feasibility study, concept design, etc. are universally used, they are frequently misunderstood. It is likely that two different engineers would define the content of each stage differently, unless they used the same common reference text. Consequently a multi-discipline project team is very likely to have mixed views on project progress. Clear project management and well-established activities for each stage in the development will readily clarify any misunderstanding. However, where these are not apparent on the project a quick estimate of the project development can be established from Table 4.1. In some instances it may be necessary to go back in the process to establish the correct design basis before going forward again.

4.3 The initial brief

The brief and design development process is essentially a methodology that converts the clients project aims into a finished building. Each new stage in the process builds on the previous stage, providing more detail of the complete solution. Consequently before starting any design work it is essential to review the documentation created in the previous phase of the development. The phase before feasibility studies is the initial brief.

The initial brief will probably contain no specific information about building services systems and may only describe the project in the broadest terms. However, a thorough initial brief will contain statements on the following broad issues:

- The purpose and scope of the project, i.e. why is it required and what it involves
- The functional requirements, e.g. process requirements, people loading, etc.
- Use of special, innovative or unusual features, e.g. high-tech building structure
- The clients' key success criteria, e.g. good people access, unique external appearance, maximum floor area, etc.
- Health, safety and environmental requirements, e.g. hazardous processes, emission constraints
- Cost, e.g. maximum capital expenditure, lowest life cycle cost
- Programme, e.g. fixed finish date or programme for lowest cost
- Quality requirements, e.g. pharmaceutical, head office, residential housing, etc.
- Aesthetic considerations, e.g. location, colour, finishes, planning requirements, etc.
- Areas of the project that require further investigation, e.g. feasibility studies

Table 4.1 Project description and early design stages

Project description	Design stage
An initial brief describes the broad project requirements. No single project solution has been agreed. Viable project options are still being discussed. Project option evaluation has not been started. Project discussions remain very global	Feasibility
A project brief describes the specific project requirements. The engineering design solution has not been agreed. Preliminary layout drawings exist although they are frequently changing. A range of engineering design solutions exist. Project discussions are focused on a single project solution. No design evaluation has occurred	Concept
A design brief describes the specific engineering requirements of the project. The design solution is known in principle. A set of concept design drawings exist, i.e. layouts, sections and engineering schematics. The main service corridors have been defined. Design value engineering is in progress. Project discussions are focused on a specific design. The engineering focus is probably on design 'production' and future information issue dates	Scheme

4.4 Feasibility studies

Feasibility studies can consider many different topics and are essentially technical 'stop or go' signals for specific project options. Subject areas that frequently involve feasibility studies are:

- Environmental impact assessments, e.g. increase traffic levels, boundary noise level restrictions, emission restrictions, etc.
- Planning permission, e.g. alternatives to plant mounted externally, site access restrictions, etc.
- Site surveys, e.g. soil conditions, services infrastructure availability, re-use of existing equipment, etc.
- Technical issues, e.g. installation of a new plant into an existing building with strict process noise limitations, design tolerance limitations, time scale issues, etc.
- Economic issues, e.g. is the cost of the design solution within the budgeted cost limit, cashflow restrictions, investment payback periods, etc.
- Building regulations, e.g. fire protection solutions for innovative designs, etc.

The study needs to determine if the project option is technically possible and whether the technical solution is consistent with the project time, cost and quality constraints. If the answer to any of these is no, the option is discarded and either the project is stopped or other acceptable project options are developed.

The majority of projects do not require feasibility studies in the strictest sense; they require a review of the proposed project options to determine their engineering implications. (A number of possible project options are usually generated during the initial brief stage through project team brain storming and general discussion.) Initially this involves qualitative engineering judgement based on past experience and preliminary calculations. These evaluation activities are carried out in conjunction with similar activities by the other members of the project team. When all the viable options have been established, the 'best fit' solution is established through a project evaluation workshop.

4.5 Project evaluation and value management

The project evaluation process is outlined in Figure 4.1. The workshop involves everybody in the project team and will probably extend beyond the clients' project manager to include other decision makers in the clients organisation.

Often the initial activities (i.e. definition of the project purpose, scope and key success criteria) have been established through the brief development process and the workshop facilitator only needs to ask the workshop to re-confirm them. Each of the key success criteria are then established as a 'need' or a 'must'. The musts are absolute project requirements, e.g. cost to be within the project budget. While the needs are more subjective project requirements where a relative judgement has to be made between the different project options, e.g. future flexibility. The needs are then prioritised by the clients' team and assigned an appropriate weighting from 1 to 10.

The viable project options that have been established through feasibility studies or engineering reviews should only need reconfirming. All this information is then gathered into a decision making matrix and scores allocated to each project option by the workshop. The scoring is based on the relative benefits of each option judged against the needs. The scoring system ranges from 10 to 1 where 10 represents an excellent match to the need and 1 represents a very poor match. The scores are multiplied by the weightings and the option with the highest resultant score is the best fit against the client's key success criteria. A sample project evaluation matrix is described in Table 4.2.

Where evaluation scores are similar, the weighting of key needs (those with the highest scores) can be incrementally adjusted in a sensitivity analysis to establish the overall effect on the total score. If two options have almost equal rankings, both may require further development until a preferred option can be determined.

All the options have to satisfy the 'must' criteria that inevitably includes ensuring the project can be procured within the budget. However, if preliminary cost information is available for each option this process can be extended to include an indication of the option value. This is achieved by dividing the option score by the option cost (i.e. the total capital and revenue cost), the highest figure represents the best value. The value figures need to be used with considerable caution as costing data at this stage in the development process is likely to be very approximate. This analysis can reveal that the 'best fit' solution does not represent the best value. In these circumstances the client, with the assistance of the project team, will need to agree the preferred project option.

4.6 Feasibility reports

Feasibility reports are usually very influential in the further development of a project and consequently need to be clear, concise and accurate. It is useful to consider how the report could be interpreted by another engineer (expert witness) in the unlikely event of a negligence claim.

A straightforward report format is:

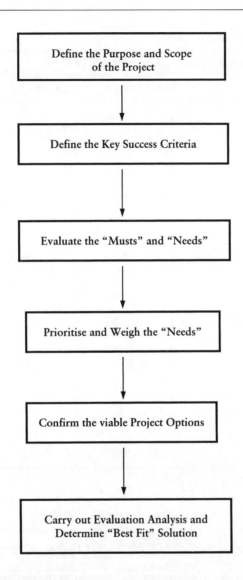

Figure 4.1 Project evaluation flow diagram.

1 Front sheet: project title, type of report, name of the author, checker and approver of the document and their signature, date the report was issued
2 Summary: no more than a single page and should include any significant engineering risks to the project
3 Table of contents
4 Definition of terms: a list of all the terms and abbreviations used in the report
5 Introduction: a couple of paragraphs describing the client, the site address, the purpose of the project and the reason for the feasibility study
6 Basis of the report: describe the assumptions and reference information used in

Table 4.2 Project evaluation matrix

Confidential project – office development		Project reference J/10295/99					
		Project option 1		Project option 2		Project option 3	
Musts							
Meets the programme requirements		Yes		Yes		Yes	
Cost within budget		Yes		Yes		Yes	
Complies with end user rental requirements		Yes		Yes		Yes	
Satisfies known planning constraints		Yes		Yes		Yes	
Satisfies corporate capital spending requirements		Yes		Yes		Yes	
	Weighting	*Score*	*Result*	*Score*	*Result*	*Score*	*Result*
Needs							
Reliability and maintainability in use	4	7	28	6	24	4	16
Innovative design solution	2	4	8	5	10	4	8
People access	8	3	24	5	40	5	40
Speed of construction	4	9	36	5	20	4	16
Provision for future expansion	4	5	20	7	28	6	24
Adequate car parking	3	2	6	8	24	8	24
External appearance	8	7	56	8	64	8	64
Maximum floor area	9	8	72	9	81	9	81
Energy efficiency	5	5	25	2	10	3	15
Total score			275		301		288

preparing the report, e.g. site survey, remaining life span of plant, inaccessible areas, etc.; this section should also describe any limitations in the use of the report

7   Description: a description of the engineering issue, site observations, technical evaluation, e.g. utility services companies reports, energy consumption comparisons, preliminary cost data, etc.

8   Discussion: an evaluation of the issues discussed in the description section, highlighting any project risk issues, any health and safety hazards and presenting fact separately from professional opinion

9   Conclusion: a list of the main report conclusions

10   Appendix:

   a   List the documents used in compiling the report, e.g. architect's layout drawings, utility services flow diagrams, current site energy usage, etc.

   b   List any reference documents used in the report, e.g. BSRIA documents, British Standards, correspondence from utility supply companies, etc.

   c   Reproduce photographs of site observations (digital cameras are particularly useful for this kind of work)

4.7 The project brief

The form and content of a project brief varies enormously and in many cases it does not exist as a specific, stand-alone document. In its simplest form it maybe a single layout drawing and brief letter; more complex projects can involve boxes of user requirement specifications, drawings, quality specifications and schedules.

The project brief is essentially a detailed development of the initial brief. The statements described in the initial brief document are expanded and defined more objectively. Additional information from feasibility studies, client reviews, project evaluation workshops, the health and safety file (established by the planning supervisor as a requirement of the CDM regulations), planning permission enquiries and contributions from specialist equipment vendors are collated into the project brief.

The brief typically contains:

- The purpose and scope of the project – repeated from the initial brief, as these should not have changed
- A description of the project based on all the current information
- Contact details of the current project team and associated groups
- Functional requirements, e.g. main function of the building, produce 150 m^3 of chemical Z per week, preliminary area footprints for the accommodation, 5,000 m^2 of storage area, design life, etc.
- Use of special, innovative or unusual features, e.g. fully integrated building management system with conditions monitoring equipment, water mist fire protection systems, performance targets, etc.
- Location of the site and access details
- The client's key success factors
- Life span of the building and its main components
- Health, safety and environmental requirements, e.g. occupation exposure level (OEL) statements for all the known products in use, energy targets, primary energy sources, CO_2 emission targets, use of non-ozone depleting refrigerants, boundary noise levels, etc.
- Cost, e.g. capital and revenue targets, budget costs for the construction work packages, cash flow projections, etc.
- Programme, e.g. design development and construction programme with some work package details
- Any specific environmental requirements, e.g. clean rooms, chilled stores, natural ventilation, internal design criteria, external design criteria, etc.
- Outline drawings, e.g. preliminary floor plans, cross sections, material flows, etc.
- Quality requirements, e.g. high care food environment requirements, reference quality standards, e.g. BS 5295: 1989 for clean environments, client installation standards, etc.
- Procurement strategy, e.g. traditional, design and build, management contract, construction management
- Aesthetic considerations, e.g. building form and construction method, anticipated foundation design, building concept impression, site layout requirements, etc.
- Details from the project evaluation and value management workshop

The project brief sets the direction for the rest of the project development through defining the preferred project option. Work during the next stage of the process should be focused on design options within the known project framework. If the project brief information fails to adequately define the project, the feasibility study and project evaluation stage will need to be repeated until a clear way forward has been established.

4.8 Actual spend and cost committed

BS 7000 Part 3 1994 Design Management Systems [1] suggests that 85 per cent of the project costs are committed by the time the project brief is written, although only 15–18 per cent of the project cost has actually been spent. This highlights the fact that the early period in the design process heavily influences the final project cost while the actual early design costs are relatively cheap. By implication, time and money spent getting it right in the early design stages is significantly more efficient than resolving design issues later in the construction process (Figure 4.2).

If the design basis is wrong or remains loosely defined until later in the project either the client requirements are not fully met or the project has to stall and go back in the design process and start again. Re-starting a project is expensive, as most costs to date are wasted and the target completion date has either to be extended or the programme accelerated to recover the lost time. In either case this usually means the project team must commit to very tight delivery programmes in order to minimise potential project delays. This inevitably promotes tension within the project as the design work progresses in parallel and time is spent producing design work rather than communicating.

4.9 Chapter review

This chapter explains the requirements of feasibility studies and project briefs through the following sections:

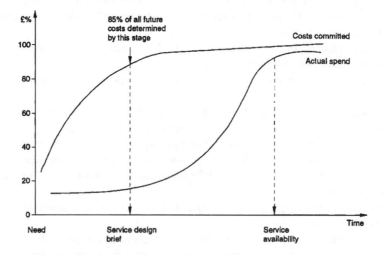

Figure 4.2 A typical curve of actual spend and cost committed. Note Figure 4.2 is Figure 3 from BS 7000, Part 3: 1994 reproduced with kind permission of BSI.

1 Establishing where you are in the development process
2 Content of the initial brief
3 Typical feasibility study issues
4 The project evaluation process
5 Format of a feasibility report
6 Content of the project brief

The next stage in the design process is concept design.

References

[1] *Design Management Systems*, BSI Publications, 1994, BS 7000: Parts 3 and 4.

Further reading

Architects Job Book, Sixth Edition, RIBA Publications, 1995.
Project Management Handbook for Building Services, Building Services Research and Information Association Publications, Application Guide AG 11/98.
A Design Briefing Manual, Building Services Research and Information Association Publications, Application Guide AG 1/90.
Design Information Flow, Building Services Research and Information Association Publications, Technical Note TN 17/92.
Value Engineering of Building Services, Building Services Research and Information Association Publications, Application Guide 15/96.

Chapter 5

Concept design stage

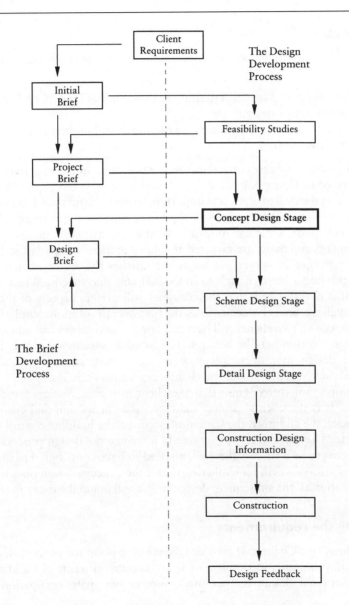

Client
Requirements

The Design
Development
Process

Initial
Brief

Feasibility Studies

Project
Brief

Concept Design Stage

Design
Brief

Scheme Design Stage

The Brief
Development
Process

Detail Design Stage

Construction Design
Information

Construction

Design Feedback

5.1 Chapter contents

The topics covered in this chapter include:

- Defining the building services requirements
- Evaluating the sources of design information
- Developing design options
- Design analysis to determine the best fit solution
- Concept design reports
- CDM requirements

5.2 Introduction

The concept design solution needs to resolve two, sometimes conflicting requirements. These are:

- Satisfy the clients key success criteria (i.e. those 'musts' and 'needs' that apply to the building services design)
- Comply with the current statutory legislation (e.g. building regulations, health and safety legislation)

The methodology for resolving these requirements is integrated into the design process described in Figure 5.1.

Before starting the design work it is important to understand that there is no perfect solution, only a solution that provides the best fit against the requirements. Perhaps an appropriate metaphor is the car industry. All the cars available meet the minimum legal requirements and many are targeted at defined markets to satisfy the demands of specific client groups. However, no single car satisfies the whole of that market as certain cars provide a better match to individual customer needs. When a customer chooses a car it is very likely that the customer still prefers aspects of the other car choices although the final choice provides the best overall solution. Similarly in building services design no two clients will have the same requirements and minor aspects of the final design solution may be better served by other system solutions. The best fit solution is a design that is compliant with the current statutory legislation and provides minimum variation from the client's key success criteria.

It is also important to recognise that the client in a construction project is not a single person. It is the clients' project manager, the clients' internal customers, the clients' financier, the architect, the structural engineer, the building control officer, the contractor, etc. They all directly or indirectly influence the design process and if the outcome is going to be a success they will all need to have contributed to the solution.

Finally, it is always worth remembering that if the concept design does not provide the best fit solution all the subsequent design stages will not make it any more suitable!

5.3 Defining the requirements

The project brief (see Chapter 4) provides the starting point for concept design work. Where this information is incomplete or is not specific in terms of building services criteria, e.g. internal design temperatures, project life span, occupation exposure

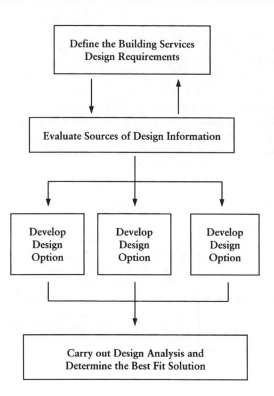

Figure 5.1 Concept flow diagram.

levels, noise levels, etc., further review will be necessary to establish the design requirements. This is an iterative process involving a review of the sources of design information and considering this information in the context of the available project information, i.e. the first two stages in the concept design flow diagram.

The evaluation process invariably generates a lot of information. A simple method of recording the data and clarifying the outcome is to consider each of the primary spaces as a box with internal environmental and process requirements and a series of inputs and exhausts. Figure 5.2 provides a generic example of this approach and can be used as a checklist during the evaluation process.

5.4 Sources of design information

A successful concept design will comply with all the necessary legal, professional and industry specific design guidance and legislation. Unfortunately there are many different sources for this type of information and each source invariably contains only a part of the overall requirements.

It is the designer's unenviable, although necessary, task to review all the applicable sources of information that impact on the design. Fortunately the magnitude of this task reduces with experience as the documentation becomes more familiar.

The common sources of design information are detailed below.

Individual Space Requirements

Outputs from the Space

Heat removal - kW
Moisture - capture velocity, condensation removal
Vapour - capture velocity m/s, hazardous zone rating
Liquids - drainage
Dust - capture velocity m/s, explosion protection
Air - m^3/s, design velocity
Noise - dB emissions

Inputs to the Space

"The Primary Space"

For Example:

Hospital Theatres
Pharmaceutical Clean Areas
Offices
Warehouse

Heat - kW process and equipment gains
Vapour - g/kg emissions
Particles - shed from people, space, process
Dust - powder handling
People - kW sensible and latent loads
Solar gain - kW transient load
Infiltration gains - internal and external

Space Criteria

Temperature - °C, set point, control band, absolute limits
Humidity - % R.H., set point, control band, absolute limits
Occupational Exposure Limits - μ/m^3 maximum and time weighted
Air changes per hour - clean up rate, industry standard, fresh air demand
Noise Criteria - NR, dB(A) environmental and work place requirements
Particle Count - B.S. or F.S. room classification
Pressure Regime - Pa, relative pressure differential, 10, 15, 25 or 50 Pa
Hazardous Area - zone rating and segregation of areas
Food Standard - high care, low care, unacceptable materials
Good Manufacturing Practice - material finishes, access

Figure 5.2 Space criteria diagram.

5.4.1 Past projects and engineering staff

The experience of other members of staff is always a useful and efficient source of information. In large organisations it is very likely that somebody will have past experience of the type of project.

When you have exhausted your known contacts within the company try placing an e-mail on the company bulletin board. A short e-mail reply with a series of reports and CAD drawings can save a lot of time and effort.

Alternatively use your own network of building services contacts.

Speak directly to the clients' end users about their problems and experiences with their current building services systems.

If your company has a formal project feedback system look up previous similar projects, speak to the engineers and site staff involved in the project. These sources of information are invaluable and can save considerable amounts of time.

5.4.2 Professional design guides and reference books

Most organisations have access to building services information databases via CD-ROM, e.g. Barbour Index. A word or subject search should provide a useful starting point in searching out more detailed information.

The surge in internet access and the vast range of information on the world-wide web make an internet search very worthwhile.

General design guidance and specific subject information is available from a range of professional design guides, reference books and supplier technical information. A broad range of useful information can be found in the publications listed in Table 5.1.

Many publications are subject to constant revision and improvement. The easiest method of ensuring your information is current is to visit the publisher's website. The following web addresses may be useful: www.cibse.org; www.ashrae.org; www.bsria.co.uk; www.bsi.org.uk (refer to British Standards section); www.hse-books.co.uk (refer to legislation section); www.ispe.org (refer to pharmaceutical industry section); www.nhsestates.gov.uk (refer to hospital work section).

5.4.3 Legislation

Engineering in the construction industry is regulated by a mountain of diverse legislation. The aim of the legislation depends on its source although in broad terms most can be placed in three categories, health and safety, quality and environment.

To assist analysis of the legislation requirements, the remainder of this section has been divided into three subsections; health and safety, building regulations and environmental protection.

5.4.3.1 Health and safety

A background to health and safety legislation and detailed information on the specific legal duties of designers is provided in Section 2.4.

The Health and Safety Commission [HSC] publish a range of excellent health and safety guidelines, derived from strategic Acts of Parliament and regulations. The impact of the guidance on building services design solutions varies. The frequently used document sections are the Approved Codes of Practice (COP and L series), Health and Safety Guidance (HS(G) series) and the Environmental Hygiene (EH series). From these sections perhaps the most frequently used documents are:

- COP 37, Safety of Pressure Systems. Approved Code of Practice, 1989
- L5, Control of Substances Hazardous to Health. Approved Code of Practice, 1999
- L8, The Prevention or Control of Legionellosis (including Legionnaires Disease)

Table 5.1 Useful publications

Publication	Subject area
Current CIBSE Guide	UK professional design guidance
Guide A	Design data
Guide B	Installation and equipment data
Guide C	Reference data
Guide D	Transportation systems
Guide E	Fire engineering
Guide F	Energy efficiency in buildings
Guide G	Public health engineering
CIBSE Applications Manuals	Comprehensive design information on specific topics. The automatic controls and condensing boilers manuals are excellent
OPUS Building Services Design File	Free to UK members of CIBSE. Provides a wide range of supplier information and contacts
Current ASHRAE Guide	American professional design guidance
Fundamentals	Comprehensive source of specific design information on a broad range of topics
HVAC Applications	Source of information for specific building services applications
Refrigeration	Same as applications although oriented towards refrigeration systems
HVAC System and Equipment	Comprehensive background information on generic plant and equipment
BSRIA Publications	BSRIA produce a wide range of excellent building services publications. This should be the starting point in any detailed information search
IOP Guide	A useful cross reference document. Mostly pipework services information
Water Supply Bye Laws Guide, 2nd Edition, ISBN 0-13-950395-1	Clear guidance on detailed water services requirements
Industrial Ventilation Manual of Recommended Practice	The design manual for industrial ventilation systems
Air Conditioning Engineering, W.P. Jones, ISBN 07131 33120	The design manual for air conditioning
Heating and Air Conditioning of Buildings, Faber and Kell, ISBN 0-7506-1858-2	Probably the most comprehensive building services textbook in the industry
Woods Practical Guide to Noise Control, Woods of Colchester	A comprehensive and user friendly industrial noise control textbook
Noise Procedure Specification, EEMUA, ISBN 0 85931 0787	A design manual for estimating off site noise levels generated by proposed industrial developments
Noise Control in Building Services, Sound Research Laboratories, ISBN 0-08-034067-9	A general textbook for noise control in building services systems
BS and BS EN Standards	The range of subject areas is huge. A subject search via one of the CD-ROM information databases is the best starting point, otherwise all the standards are listed in the back of OPUS
Engineering for Food Safety and Sanitation, T.J. Imholte, ISBN 0-918351-00-6	A general food engineering safety book with useful information for building services applications in the food industry
Cleanroom Design, W. White, ISBN 0-471-92814-3	A comprehensive textbook detailing all aspects of cleanroom design and specialist water services

Table 5.1 (*continued*)

Publication	Subject area
An Introduction to Contamination Control, W. White	A seminar booklet providing an overview of contamination control in clean environments
Cleanroom Technology, W. White	A seminar booklet introducing cleanroom technology
LPC Rules for Automatic Sprinkler Installations	The UK insurers standard rules for sprinkler installations. Incorporates BS 5306 Pt 2:1990 and all current technical bulletins. Unfortunately the BS information flow is difficult to follow
FM, Loss Prevention Data	FM publish data sheets and technical advisory bulletins on a wide range of subjects all relating to loss control. They are the American insurers of standard design requirement
Guide to the Selection and Installation of Compressed Air Services, British Compressed Air Society	An introduction to compressed air systems
Suppliers Data	Many suppliers provide very good technical support information. Typically, Spirex Sarco for steam and compressed air equipment, Camfil for filters and Woods for noise control equipment and smoke control

Approved Code of Practice, 1995

- L24, Workplace Health, Safety and Welfare. Approved Code of Practice, 1996
- HS(G) 37 Introduction to Local Exhaust Ventilation, 1993
- HS(G) 39 Compressed Air Safety, 1998
- HS(G) 50 The Storage of Flammable Liquids in Tanks, 1998
- HS(G) 51 The Storage of Flammable Liquids in Containers, 1998
- HS(G) 54 The Maintenance Examination and Testing of Local Exhaust Ventilation, 1990
- HS(G) 56 Noise at Work. Noise Assessment, Information and Control. Noise Guides 3–8, 1990
- HS(G) 70 The Control of Legionellosis including Legionnaires Disease, 1993
- HS(G) 103 Safe Handling of Combustible Dusts, 1994
- EH 40/99 Occupational Exposure Limits

Note: EH 40 is issued annually with a corresponding upgrade to the extension, e.g. EH 40/2000, etc. Unusually it is directly referenced in the Control of Substances Hazardous to Health Regulations 1994 and consequently has the same legal stature as the regulations.

The building services requirements outlined in the health and safety documentation are generally descriptive, i.e. a performance requirement, without stating how it should be achieved. 'How' the performance is achieved is the duty of the 'designer'.

Techniques for reducing health and safety risks have a prescribed hierarchy ranging from a safe place (the best solution) to a safe person (the poorest solution). The approach is described in Figure 5.3.

Generally the cost of providing a pair of safety glasses is less than the cost of

	Technique	Description
Safe Place	Substitute or Eliminate	Eliminate the risk, e.g. take the solvents out of the process, don't provide cooling towers.
	Reduce	Reduce the risk, e.g. reduce the use of solvents in the process to a level that represents minimum risk, ensure spare parts can be safely lifted by one person.
	Enclose or Isolate	Separate the risk from the person, e.g. place the noisy equipment in a room that isn't a workplace, use a safety cabinet.
	Control	Control the risk, e.g. provide dilution ventilation, reduce the exposure time, provide a safe system of work.
	Protection	Provide protection to the person, e.g. PPE, breathing air, safety glasses.
Safe Person	Discipline	Supervisory control, e.g. safety signs.

Figure 5.3 Health and safety risk control hierarchy and techniques.

eliminating the risk. However, the design solution is determined by a 'reasonably practicable' approach. This means that a balance must be made between the cost of the works and the control of the health and safety risk. This can be a difficult judgement to make, although the vast majority of clients understand their health and safety obligations and start with elimination rather than protection.

A general example of a reasonably practicable approach could be the dispensing of powders with an OEL of 100 μ/m^3. This could be safely carried out in either a downflow containment booth by a person with a face mask or with a local extract ventilation (LEV) system and a person fully clad with a breathing air system. The containment booth control solution will be significantly more expensive than the person protection solution. If the powder dispensing process was part of a new £20 million project and dispensing occurred five or six times every day, a reasonably practicable solution would be the containment booth. If, on the other hand, the project budget was £6,000 and powder dispensing only occurred once a month for no more than 5 minutes, then personnel protection may be acceptable.

5.4.3.2 Building regulations

The building regulations apply to all new project work in England and Wales (similar regulations apply in Scotland). They are prescriptive requirements that are updated to incorporate feedback from the existing building stock, fire incidents and mounting environmental concerns, e.g. energy efficiency.

The regulations are explained in a set of approved documents distributed through HMSO. The approved documents require careful study as they contain significant amounts of detailed information and frequently cross-reference with other documentation, e.g. British Standards.

The main sections of the building regulations that directly effect building services design are listed below. *This guidance must be read in conjunction with the approved documents.*

Approved document B, fire safety

B3: internal fire spread (structure), provides guidance on compartment sizes and the requirement for sprinkler protection. (*Note*: compartment size 'trade-offs' against sprinkler protection require a BS 5306: Part 2 life safety sprinkler system, effectively an LPC sprinkler system with additional requirements.)

B5: access and facilities for the fire service, highlights requirements for the provision of fire fighting shafts.

Approved document E, resistance to the passage of sound

Outlines typical constructions that restrict the passage of airborne and impact sound within dwellings.

Approved document F, ventilation

F1: means of ventilation, defines the minimum ventilation requirements in domestic and non-domestic applications.

Approved document G, hygiene

G3: hot water storage, provides guidance on unvented hot water storage systems.

Approved document H, drainage and waste disposal

H1: sanitary pipework and drainage, explains the minimum requirements for sanitary pipework and foul drainage.

Approved document J, heat producing appliances

J1/2/3: heat producing appliances, defines the requirements for supply air and flues from heat producing equipment.

Approved document L, conservation of fuel and power

L1: conservation of fuel and power, is primarily known for stipulating the elemental building fabric 'U' values (currently roofs = 0.25 W/m^{2}K with exceptions, exposed walls = 0.45 W/m^{2}K, windows = 3.3 W/m^{2}K, floors = 0.45 W/m^{2}K). It also provides information on space heating, hot water storage controls and insulation.

5.4.3.3 Environmental protection

The Environmental Protection Act 1990 (EPA) aims to 'prevent, minimise and render harmless' the emissions of 'prescribed processes and substances'. The EPA enforcing authority is currently split between the environmental agency (Part A: prescribed processes and prescribed substances released to air, water and land) and the local authority (Part B: prescribed processes and prescribed substances released to air). Background to the EPA and the prescribed processes and substances can be found in the National Society for Clean Air and Environmental Protection, Pollution Handbook [1], which is reprinted every year.

Typical environmental protection issues that can involve building services design work include:

- Containment of spillages and fire water run-off
- Volatile organic compound (VOC) extract systems
- Stack dispersion modelling
- Sulphur dioxide and nitrogen oxide emissions
- Environmental noise (see below)

The general approach to resolving environmental engineering issues is described as BATNEEC or best available techniques not entailing excessive cost. The methodology behind BATNEEC is described in Ref. [1] and is similar to a 'reasonably practicable' approach, in that it involves a judgement balancing a technical solution and its cost.

Environmental noise problems provide an exception to the general BATNEEC approach. They are categorised as statutory nuisances and require a solution independently of almost any cost restraint.

The inclusion of noise as one of a number of specific statutory nuisances (others include smoke, gases, etc.) within the EPA is expressed in two separate definitions:

Noise emitted from premises so as to be prejudicial to health or a nuisance

and

Noise that is prejudicial to health or a nuisance and is emitted from or caused by a vehicle, machinery or equipment in a street

The local authority has a duty to ensure that their area is regularly inspected to detect if a statutory nuisance exists or is likely to occur or recur. This duty in the case of environmental noise will probably be devolved to the environmental health officer (EHO). Likewise the local authorities duty to take such steps as are reasonably practicable to investigate any compliant of statutory nuisance from a person living in the area will pass to the EHO. Practically, this means if a single person living in the area of the new works complains the EHO will investigate. The EHO only has to ascertain that a nuisance exists before issuing an abatement notice. The notice may impose all or any of the following restrictions: (a) abatement of the nuisance or prohibiting or restricting its occurrence or recurrence; (b) the carrying out of such works and other steps necessary to abate the nuisance. The consequence of an abatement notice being served on the works could be a permanent shutdown of the facility and a total loss of business.

On new works the environmental noise design requirements are understandably imposed by the local authority as part of the planning permission consent. In recent years the planning consent terms have tended towards 'no increase in existing background noise level'. (This prevents a gradual increase in background noise because if the new works matched the existing level the resultant addition would cause a 3 dB increase.) Consequently if no planning permission information is available at the concept design stage the basis of design can fairly confidently be taken as 'no increase in existing background level'. This is equates to a design basis of -10 dB beneath the existing background level.

The background noise level is typically established by recording the L_{90} dB level (i.e. the level exceeded 90% of the time across an octave band spectrum from 31.5 Hz to 16 kHz) between 02:00 and 04:00 h on weekday mornings at an agreed reference location(s).

Note: the environmental regulations in the UK are set to change. The Environment Act 1995 made way for the integrated pollution prevention and control (IPPC) regulations that come into force in November 1999. These regulations will change how a large part of industry is regulated, e.g. BATNEEC is replaced by best available technique (BAT). It may also alter the way that environmental noise is regulated as responsibility may move towards the EA and away from the local authority. IPPC regulations also include energy efficiency requirements.

5.4.4 Industry specific design information

5.4.4.1 Clients' design standards

Many clients, e.g. Shell, Marks and Spencer, etc., have their own design standards. American clients tend to use factory mutual (FM) design standards as these describe the insurers' minimum requirements. Client specific design standards need to be formally requested during the concept design stage. This ensures that either they do not exist or that where they are available they are incorporated into the concept design.

5.4.4.2 Pharmaceutical industry

Design issues in the pharmaceutical industry can be broadly summarised in the expression 'protect the people from the product and the product from the people'. Protecting the people from the product or risk is the object of the previously described health and safety legislation. The uniqueness of pharmaceutical project work is protecting the product from the people or ensuring product quality. This issue in the UK, in addition to the obvious commercial necessity, is regulated through the Medicines Act (similar legislation exists in America and Europe). The effect of this legislation on building services engineering design is a consequence of quality assurance. This all encompassing issue involves the application of good manufacturing practice (GMP) and validation.

GMP process requirements involve the integration of the process into the building facility. A GMP review would investigate issues such as good materials flow, no cross-contamination, material finishes, etc. The complete solution to this type of issue is unique to the specific process and the individual building facility. Consequently, pharmaceutical design information and standards provide little objective information on GMP. However, the detail of the solution could typically involve the following requirements:

- Pressure regimes between product and non-product areas
- Pressure regimes within product areas
- Containment or enclosure of product transfer points, e.g. dispensing booths, closed transfer couplings

- Classified, clean product areas, e.g. BS 5295
- HEPA supply and extract filtration
- Clean, non-shedding surface finishes, e.g. stainless steel rather than galvanised ductwork
- Fixings that are easy to clean and that minimise contamination risk, e.g. no slotted head screws or pop rivets
- Joints that are easy to clean and that minimise contamination risk, e.g. welds ground smooth
- Minimum of exposed services in the product areas
- Easy clean production surfaces, e.g. curved radius corners inside isolators
- Segregation of plant and equipment serving product and non-product areas

Validation of the building services systems primarily affects those systems that involve product contact. The design solution has to be robust enough to consistently maintain the design conditions and be capable of proving this requirement. It is the conditions that have to be validated not necessarily the means of achieving them. This may mean, for example, that it is more beneficial to use an independent monitoring system to sample the space conditions rather than attempt to validate the whole of the building services BMS or control system. Common reference standards for building services design in the pharmaceutical industry are:

- Food and Drug Administration (FDA) GMP guide – necessary for all companies that want to retail their products in the US market
- 'Orange Guide', Rules and Guidance for Pharmaceutical Manufacturers 1997 incorporates the EC directives on GMP and is valid for all companies that want to retail their products in the EU
- Federal Standard 209E, Cleanroom and Work Station Requirements, Controlled Environment generally applies to all projects that manufacture for the US market
- BS 5295: 1989 Environmental Cleanliness in Enclosed Spaces:
  Part 0: General introduction, terms and definitions for clean rooms and clean air devices
  Part 2: Method for specifying the design, construction and commissioning of clean rooms and clean air devices
  Part 3: Guide to operational procedures and disciplines applicable to clean rooms and clean devices
  Generally applies to all projects that manufacture for the UK market
- International Society of Pharmaceutical Engineering (ISPE), Baseline Guides – these guides are a joint production with the FDA and will no doubt become the industry de-facto standards; the current ten guide series is listed below, however, many guides are still in draft form and remain unpublished (the published guides date of issue are marked in brackets; the web site address is www.ispe.org)

  – Biotechnology
  – Bulk Pharmaceutical Chemicals (published June 1996)
  – Medical Devices
  – Oral Solid Dosage Forms (published March 1998)
  – Oral Liquids and Aerosols
  – Packaging and Warehousing

- R&D Facilities
- Sterile Manufacturing Facilities (published February 1999)
- Water and Steam Systems
- Commissioning and Qualification

5.4.4.3 Food industry

The food industry is not currently as regulated as the pharmaceutical industry and consequently the amount of authoritative design guidance is limited. However, the approach to establishing a design concept is similar to the pharmaceutical GMP review and the detail in the solutions is similar although not as onerous. Typical design detail could include:

- Ensuring air movement from clean high risk areas to less clean low risk areas
- Pipe runs free from dead legs
- Mounting plant, equipment and services off the floor and away from walls to allow easy cleaning
- Minimum overhead services that collect dust and trap bugs that can fall onto the product
- Minimum number of flange joints. No spiral wound ductwork as joints trap dust
- Minimum number of fasteners and exposed threads
- No glass or painted products

Frequently food manufacturer's design requirements are driven by their clients, e.g. Marks and Spencer or Sainsburys. Where these guidelines are available they should be used, otherwise the best generic guidance available is published by the Campden Food and Drink Research Association. Their publications include:

- Guidelines on Air Quality Standards for the Food Industry (1996)
- Technical Manual 44, Design and Construction of Walls, Ceilings and Services for Food Production Areas (1994)

In recent years there have been a number of 'food scares'. Part of the UK government response is the creation of a new Food Standards Agency. The aims of the new body are:

- Monitor the safety and standards of all food for human consumption
- Commission scientific research and develop new policies
- Co-ordinate and monitor the standards of food law enforcement
- Advise the public, Ministers and the food industry

Consequently it is likely the agency will issue new guidance on food hygiene standards that will supplement the Food Supply (General Food Hygiene) Regulations 1995.

5.4.4.4 Hospital work

Work in the UK on National Health Services (NHS) hospitals is based on its own specific set of design guidance. The guidance is very comprehensive and detailed. In many cases a specific type of building services system or specialist department may involve the review of a number of documents. The main document categories are:

- Design guides
- Fire code
- Health building notes (HBN)
- Health technical memoranda (HTM)
- Model engineering specification
- Health facilities notes (HFN)
- Health guidance notes (HGN)

Up to date lists of publications and a useful word search engine are available on the NHS estates web site at www.nhsestates.gov.uk. If you do not have access to the documents on CD-ROM they can all be purchased through the stationary office.

5.4.5 British Standards

British Standards provide a very wide range of detailed information. The majority of the information is specific to installation requirements, i.e. BS 4504:1989 circular flanges for pipes, valves and fittings. However a number of the standards do provide design guidance that could impact on concept work.

British Standards that contain building services design guidance are listed below. This is not a complete list and this information must be read in conjunction with a general review of the British Standard publications.

5.4.5.1 Heating, ventilation and air conditioning

- BS 4434 Specification for Safety and Environmental Aspects in the Design, Construction and Installation of Refrigerating Appliances and Systems, 1995.
- BS 5295 Environmental Cleanliness in Enclosed Spaces, Pt. 0, Pt. 2 and Pt. 3, 1989 (as above).
- BS 5572 Code of Practice for Sanitary Pipework, 1994.
- BS 5720 Code of Practice for Mechanical Ventilation and Air Conditioning in Buildings, 1979.
- BS 5726 Microbiological Safety Cabinets, Parts 1–4, 1992.
- BS 5918 Code of Practice for Solar Heating Systems for Domestic Hot Water, 1989.
- BS 5925 Code of Practice for Ventilation Principles and Designing for Natural Ventilation, 1991.
- BS 6230 Specification for Installation of Gas Fired Forced Convection Air Heaters for Commercial and Industrial Spaces, 1991.
- BS 6644 Specification for Installation of Gas Fired Hot Water Boilers of Rated Inputs Between 60 kW and 2 MW, 1991.
- BS 6700 Specification for the Design, Installation, Testing and Maintenance of Services Supplying Water for Domestic Use within Buildings and their Curtilages, 1997.
- BS 6880 Code of Practice for Low Temperature Hot Water Heating Systems of Output Greater than 45 kW, Parts 1–3, 1988.
- BS 7258 Laboratory Fume Cabinets, Parts. 1–4, 1994.

- BS 8301 Code of Practice for Building Drainage, 1991.
- BS 8313 Code of Practice for Accommodation of Building Services in Ducts, 1997.

5.4.5.2 *Fire protection*

- BS 5306 Fire Extinguishing Installations
  Part 0: Guide for the Selection of Installed Systems and other Fire Equipment, 1986
  Part 1: Hydrant Systems, Hose Reels and Foam Inlets, 1976
  Part 2: Specification for Sprinkler Systems, 1990 (to be read in conjunction with the LPC rules for automatic sprinkler installations)
  Part 3: Code of Practice for Selection, Installation and Maintenance of Portable Fire Extinguishers, 1985
  Part 4: Specification for Carbon Dioxide Systems, 1986
  Part 6: Foam Systems
  Section 6.1: Specification for Low Expansion Foam Systems, 1988
  Section 6.2: Specification for Medium and High Expansion Foam Systems, 1989
  Part 7: Specification for Powder Systems, 1988.
- BS 5588: Fire Precautions in the Design, Construction and use of Buildings
  Part 0: Guide to Fire Safety Codes of Practice for Particular Premises/Applications, 1996
  Part 1: Code of Practice for Residential Buildings, 1990
  Part 4: Code of Practice for Smoke Control using Pressure Differentials, 1988
  Part 5: Code of Practice for Fire Fighting Stairs and Lifts, 1991
  Part 6: Code of Practice for Places of Assembly, 1991
  Part 7: Code of Practice for the Incorporation of Atria in Buildings, 1997
  Part 9: Code of Practice for Ventilation and Air Conditioning Ductwork, 1999
  Part 10: Code of Practice for Shopping Complexes, 1991
  Part 11: Code of Practice for Shops, Offices, Industrial Storage and other Similar Buildings, 1997
- BS 5908: Code of Practice for Fire Precautions in the Chemical and Allied Industries, 1990.

British Standards are produced by the British Standards Institution (BSI) and provide guidance on all aspects of engineering work. Harmonisation of standards across Europe has led to the development of BS EN documents, i.e. English language versions of agreed European Standards. Increasingly BS EN will replace BS as the European market harmonises.

A complete list of all current BS and BS EN is detailed in the Building Services OPUS Design file, issued annually to members of CIBSE. Alternatively, visit the British Standards Institution website at www.bsi.org.uk.

5.5 Hierarchy of legislation

The legal hierarchy of documentation that applies to any project subject to British law is outlined in Table 5.2. The hierarchy descends in order of priority from Acts of

Parliament down to HSC guidance notes and British Standards. It is unlikely that an Act of Parliament will provide any objective design information so the first practical design documents are likely to be the regulations.

It is essential that any design solution satisfies the legal requirements imposed by statute. This is not simply because failure to comply with the legal requirements could result in prosecution and have serious commercial consequences. The legal requirements strive to create a safe environment for employees and the public. Non-compliant designs can have fatal consequences, e.g. failure of a smoke ventilation system in an emergency.

It is also important to be able to separate the legal requirements from the contractual requirements. The contractual requirements to either 'exercise reasonable standards of professional care and skill in carrying out the service' or 'provide services that are fit for purpose' (refer to Section 3.3) impose additional requirements. It would be very easy to provide a design solution that met all the legal requirements that was not fit for purpose. The design solution needs to comply with both sets of requirements otherwise the design could be safe but potentially useless. Contractual requirements in a hospital design may mean complying with the current health technical memoranda (HTMs), in a pharmaceutical design complying with current GMP.

In many situations the contractual requirements exceed the legal obligations in which case the more onerous requirements apply. However, conflicts in requirements do occur and they can be awkward to resolve amicably.

Table 5.2 The legal hierarchy of information

Type of documentation	Priority	Example
Act of Parliament	I	Health and Safety at Work Act
		Environment Protection Act
		Medicines Act
Regulations	2	COSHH Regulations
		Building Regulations
		CDM Regulations
		Noise at work regulations
HSC approved codes of practise	3	L8, The Prevention or Control of Legionellosis
		COP 37, Safety of Pressure Systems
HSC guidance notes	4	All other HSE documents
		HS(G) 37 Introduction to Local Exhaust Ventilation
		HS(G) 51 The Storage of Flammable Liquids in Containers
British Standards	5	BS 5544 1991 Fire Precautions in the Design and Construction of Buildings
Professional guidance		CIBSE Guides
		Technical Memorandum and Application Manuals
		ASHRAE Guides
		BSRIA Information
Client or industry standards		Orange Guide
		Baseline Guides
		NHS Estates Guidance

5.6 Develop the design options

Having established the design requirements and researched the appropriate sources of design information, it is likely that a number of 'possible' design solutions will present themselves. Straightforward work, e.g. an extension to a sprinkler system, a small factory unit or a small hotel, etc. may have an obvious design solution, however, most project work will require further design evaluation. Preliminary investigation for a new office block may well have established that the building requires comfort cooling although no single design may appear to represent the best solution. The aim of this stage in the process is to develop three or four design alternatives that all meet the broad requirements of the project. The office project, for example, may consider fan coils, chilled beams, VAV, natural ventilation and pre-cooling, solutions.

The design alternatives should be developed as broad concept schematics and, where appropriate, as typical layout sketches. This type of information will allow the concept design requirements to be clearly established and will also provide an unambiguous method of communication.

5.7 Evaluate the options and determine the best fit solution

When the concept options have been established they need to be evaluated against the appropriate clients' key success criteria defined in the project brief and any other project specific building services criteria. The specific success criteria will vary on every project although typically the following may be considered:

- Cost. What are the capital and revenue costs? How do the costs compare with the project brief requirements?
- Programme. Can the design, installation and commissioning meet the programme restraints?
- Constructability. Can the design be practically constructed without significant safety risk? What effect may the installation have on the continuous operation of the facility? Are there any weight restrictions on the structure? Are there any fixing restrictions on the structure? Are there any physical access restrictions? Are any of the buildings listed? Are there any planning restraints that affect the building services systems? What techniques are integrated into the design solution to maximise site productivity?
- Reliability. How does the design satisfy the clients production or retail demands? How has the design considered security of supply? What provision has been made for duty/standby plant and equipment?
- Future demand. What provision has been made for future demand? How will the future demand be integrated into the current design solution?
- Internal environment. How does the design satisfy the internal environmental criteria? How does the design satisfy the GMP and food hygiene requirements? How does the design respond to variations in load? How is the pressure regime maintained?
- Plant space. What are the proposed services plant room space requirements? How will the services be distributed through the building? How will plant and equipment be installed? How does the design solution load the building? How do you access the plant rooms after construction? What lifting provision needs to be provided?

- Maintainability. What access provision has been made for replacing plant and equipment? What provision has been made for maintenance shutdowns?
- Sustainability. What are the main fuel sources? Does the project require a BREEAM assessment? How does the design provide an energy efficient solution? What are the likely carbon dioxide, sulphur dioxide and nitrogen oxides emissions? Does the design solution utilise any renewable energy sources?
- Innovation/complexity. How complex is the design solution? Does it involve numerous specialist contractors? Is the design solution unique? Is the technology tried and tested?

The list is not inclusive and any relevant topic that has been highlighted as important to the project success should be added.

Each of the concept design solutions are then evaluated against the success criteria using a decision matrix (see Section 4.5). All the criteria should be considered as 'needs' as no concept should be put forward that does not satisfy the project 'musts'. Ideally the evaluation process will involve all the key people contributing to the building services solution, e.g. the clients' project manager, the project managers' internal clients, the quantity surveyor, the architect, the contractor (if they are involved in the project at this stage). This approach may not be practical or the project team may only expect to be informed of the solution and justification process. In these situations an internal group of 'design peers' should be asked to review the proposed solutions and contribute to the design evaluation process.

The scoring is based on the relative benefits of each option judged against the criteria. The criteria are prioritised by the group and assigned an appropriate weighting from 1 to 10. The scoring of the options against the criteria also ranges from 1 to 10 where 10 represents an excellent match to the criteria and 1 represents a very poor match. The criteria scores are multiplied by the weightings and totalled. The concept design with the highest resultant score is the best fit solution against the evaluation criteria.

A concept design evaluation matrix is described in Table 5.3.

The evaluation criteria are based on the office development project evaluation matrix provided in Table 4.2 and a number of possible project specific building services criteria. Before recommending option 3 (the design option with the highest score) it would be prudent to carry out some simple sensitivity analysis checks to determine the overall effect on the decision analysis.

A measure of value can be introduced into the evaluation process by dividing the option score by the option cost (i.e. the total capital and revenue cost). The option with the highest score represents the best value. This approach would be very useful where accurate option costs were known (in other circumstances it could be misleading). In these situations the capital and revenue costs would be taken out of the 'needs' criteria as they would not require a subjective assessment.

The aspects of the best fit design solution that involve building control (fire officer), HSE, EHO, etc. will need to be discussed and 'closed out' with them during the concept design stage to ensure that the risks to the design are minimised.

Table 5.3 Concept design evaluation matrix

Confidential project – office development, concept design options		Project reference J/10295/99					
Needs	Weighting	Design option 1		Design option 2		Design option 3	
		Score	Result	Score	Result	Score	Result
Capital cost	9	7	63	9	81	8	72
Revenue cost	7	6	42	4	28	7	49
Programme duration	8	7	56	7	56	8	64
Reliability and maintainability in use	6	7	42	5	30	6	36
Innovative design solution	4	4	16	6	24	8	32
Maximum floor area	9	7	63	8	72	7	63
Energy efficiency	7	6	42	5	35	6	42
Provision for future expansion	3	6	18	6	18	5	15
Speed of construction	7	7	49	7	49	7	49
External plant requirements	6	5	30	6	36	5	30
Zone control	5	8	40	7	35	7	35
Total score			461		464		487

5.8 Concept design report

The concept design solution is the basis for the future design development and needs to be clearly communicated to the rest of the project team.

Where the project team have been involved in the design evaluation process the concept schematics and design basis will have been presented at the evaluation workshop. In this situation it may only be necessary to circulate the meeting notes, decision matrix and proposed design schematics to the project team. Alternatively if the project team have not been involved in the decision analysis, or where the project team is very large and only a small section of the team have been involved in the workshop, the design analysis results will need to be presented to the team and supported with the concept design report.

The concept design report should supplement and develop any project brief and feasibility information without becoming a technical design document (see Chapter 6). The report should convey the following information:

• The reasoning behind the design selection, i.e. the choice of evaluation criteria
• The range of design concepts considered during this stage of the design process
• The design solution that is going to be developed during scheme design

The report format will need to suit the project requirements; however, a typical format is:

• Front sheet: project title, type of report, name of the person who prepared, checked and approved it and their signature, date the report was issued and the revision number

- Summary: no more than a single page and should include any significant engineering risks to the project
- Table of contents
- Definition of terms: a list of all the terms and abbreviations used in the report
- Introduction: a couple of paragraphs describing the client, the site address, the purpose of the project, the reason for the concept design report and the sources of information used in preparing the report
- Description: a description and explanation of the evaluation criteria, a description of the proposed concept design solutions with cross referencing to the concept schematics in the appendix
- Discussion: an evaluation of the issues discussed in the description section, highlighting the advantages and disadvantages of each concept design.
- Conclusion: the decision analysis matrix, an outline technical description of the agreed design solution including any risks it presents to the project, a statement confirming the way forward
- Appendix
  - Schematics of the concept design proposals
  - Any reference documents used in the report to explain the technical issues and support the conclusion, e.g. BSRIA documents, British Standards, current technical and trade publications
  - Initial feedback from building control (fire officer), HSE, EHO, etc.

5.9 CDM requirements

The main requirements of the regulations are described in Chapter 2.

The Construction (Design and Management) Regulations 1994 impose a number of duties on designers [1]. The duties that directly affect the design work are:

- identify the significant health and safety hazards and risks of any design work
- give adequate regard to the hierarchy of risk control (see Figure 5.3)

Most companies involved in design have developed their own methods and documentation to ensure they comply with the CDM regulations [2]. Where this type of system is not available an alternative approach would be to work through the following process.

5.9.1 CDM *design review process*

1 Define the building services scope of work and the physical boundary limits of the new services
2 Review the health and safety hazard checklist of topics detailed in the next section
3 Identify those issues derived from the checklist that represent a significant health and safety hazard on the project; 'significant' hazards are generally identified through a risk assessment process (see Chapter 7), however, at this stage in the design development process it is more practical to list all the potential hazards as the level of design detail is probably not adequate to carry out a meaningful risk assessment

4 Record the project list of hazards
5 Review the hazards against the health and safety risk control hierarchy
6 Record how the project risks are going to be controlled and ensure these actions are integrated into the design solution

5.9.2 Health and safety hazard check list

Does the project involve any of the following hazards.

Design

- Use of relevant fluids, as defined in the pressure systems and transportable gas containers regulations, e.g. steam, compressed air, nitrogen, helium, hydrogen, etc.
- Fire protection, e.g. sprinkler systems, deluge systems, staircase pressurisation, hydrants, hose reels, wet or dry risers, etc. Define the known extent of the mechanical fire protection systems and areas for future investigation
- Fire compartmentation, e.g. fire dampers, fire stopping, fire rated ductwork, etc. to maintain the fire compartmentation. Is there a requirement for smoke extract ventilation, smoke curtains, etc.?
- Design of systems within hazardous areas or to control hazardous area environments, e.g. flammable hazard zones 0, 1, 2 and dust hazard zones X and Y
- Legionella, e.g. hot water storage, potable water distribution, cooling towers, etc.
- Exposed hot surfaces above 55°C, e.g. radiators, steam traps, boilers, etc.
- Pressure relief valves (PRVs), e.g. safe discharge location for LTHW, MTHW, steam systems, etc. Vacuum relief on vacuum systems, etc.
- Containment of hazardous solids and liquids, e.g. dispensing and transfer of low OEL solids with a pharmaceutical facility
- Dust explosion, e.g. dust extract systems and explosion venting, suppression or containment.
- Specialist gas extract systems, e.g. extract from solvent laden processes, specialist laboratory and hospital theatre extracts, etc.
- Noise control, e.g. are there any areas of the design that require investigation to ensure they comply with the noise at work regulations? Are there any systems, plant or equipment that may create a statutory nuisance under the EPA?
- Refrigerant discharge, e.g. could the rupture of a refrigerant distribution line within an enclosed space lead to the build-up of a hazardous environment?
- Future maintenance requirements, e.g. safe routes into plant areas for people and materials, adequate head height, replacement of motors, tube bundles, coils, filters, boiler parts, access to fire dampers, valves, etc.
- System isolation, e.g. emergency isolation of gas services, double isolation valves and drain on steam supplies to segregate the point of access and the live steam, adequate section isolation valves, adequate standby plant and equipment to allow continuous use and maintenance shutdown, etc.
- CE marking, e.g. who is responsible for CE marking the installed systems?

Construction

- Disruption of hazardous materials, e.g. asbestos insulation, lead pipework, fibrous materials, contaminated waste material
- Abnormal demolition work, e.g. strip-out of contaminated ductwork or services in hazardous areas
- Breaking into existing systems, e.g. steam mains, compressed air, sprinkler systems, fire alarms, etc.
- Work in restricted spaces, e.g. trenches, vessels, basements, etc.
- Work in hazardous areas, e.g. sources of ignition in flammable atmospheres, hazardous liquids and vapours, etc.
- Abnormal manual handling requirements, e.g. the physical transportation of AHU flat packs through confined spaces, general manual handling issues
- Abnormal cranage of plant and equipment, e.g. lifting of AHUs onto a roof involving road closures
- Significant imposed loads on the building structure, e.g. 24 h water storage tanks on the roof, adequate structural supports to large pipework mains
- Work involving noisy processes, e.g. new work within existing noisy production areas, work associated with demolition
- Maintenance access involving roofs, e.g. fragile roof materials, lighting, handrails, access and egress for materials, etc.
- Hazardous construction materials, e.g. solvents, fibrous materials, irritants, etc.
- Working at height, e.g. safe scaffolding, use of mobile towers, etc.
- site fabrication, e.g. cutting, welding, pressure testing, etc.

5.10 Chapter review

This chapter describes the requirements of concept design work through the following topics:

1 Defining the requirements
2 Sources of design information:
   - Past projects
   - Design guidance
   - Legislation
   - Industry specific design information
   - British Standards
3 Hierarchy of legislation
4 Developing the design options
5 Evaluating the design options
6 Concept design reports
7 CDM requirements:
   - CDM design review process
   - Health and safety hazard check list

After the concept design stage the design development process stops while the future design basis is defined in the design brief stage.

References

[1] *A Guide to Managing Health and Safety in Construction*, HSE Books.
[2] *Designing for Health and Safety in Construction*, HSE Books.

Further reading

Architects Job Book, Sixth Edition 1995, RIBA Publications.
A Design Briefing Manual Building Services Research and Information Association Publications, Application Guide AG 1/90.
Design Information Flow, Building Services Research and Information Association Publications, Technical Note TN 17/92.
Value Engineering of Building Services, Building Services Research and Information Association Publications, Application Guide 15/96.
Design Management Systems, BSI Publications, 1994, BS 7000: Parts 3 and 4.
Building Services Legislation, Building Services Research and Information Association Publications, Reading Guide 14/95.

Chapter 6

The design brief

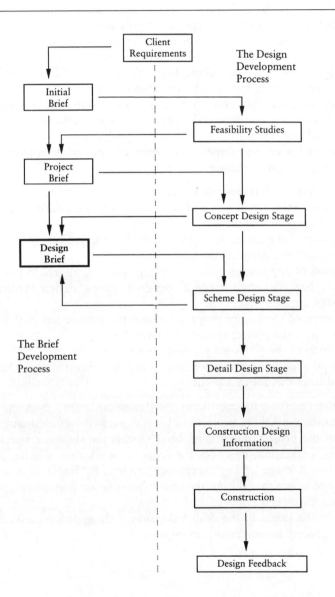

6.1 Chapter contents

The following topics are covered in this chapter:

- The function of a design brief
- Contents of a design brief

6.2 The function of a design brief

Concept design involved the review of significant amounts of documentation; it generated many ideas, it probably left a number of good ideas stored for future use and ended with the agreement of a concept design solution.

The knowledge and work that went into creating the concept design solution is probably scattered across the project files, site survey notes, thoughts from informal discussions, file notes, copies of technical reports, etc. and held together loosely in the head of the designer ready for the start of scheme design!

Rather than launch straight into scheme design the design brief provides an essential opportunity to consider and record the previous design work, plan the future design development and communicate the issues to the project team.

The design brief document satisfies important design and design verification functions. The design functions include:

- Clarification of the future scope of services
- Clarification of the actual scope of work (based on the concept design)
- Confirmation of the technical basis of design
- Confirmation of the design standards applied to the project
- Clarification of the design deliverables
- Clarification of any unique project requirements, e.g. approach to solving health and safety hazards, environmental design features, design margins, technical design risks, etc.
- Confirmation of the design programme and the integration of the design work with the rest of the project team activities.
- Confirmation of the proposed procurement route
- Creation of a design 'yardstick': any future significant design changes can be measured from this starting point

The design verification function links the design audit trail from the initial brief, project evaluation workshop, project brief and concept design evaluation through to completion of the project. The design brief repeats the success criteria used in the concept design evaluation, describes the scope of work that is going to satisfy the criteria and lists a range of key acceptance criteria. By implication when the key acceptance criteria are met during the commissioning or post-completion stage the success criteria have been achieved.

Although all the stages in the design decision making process have not yet been discussed the process is summarised in Figure 6.1.

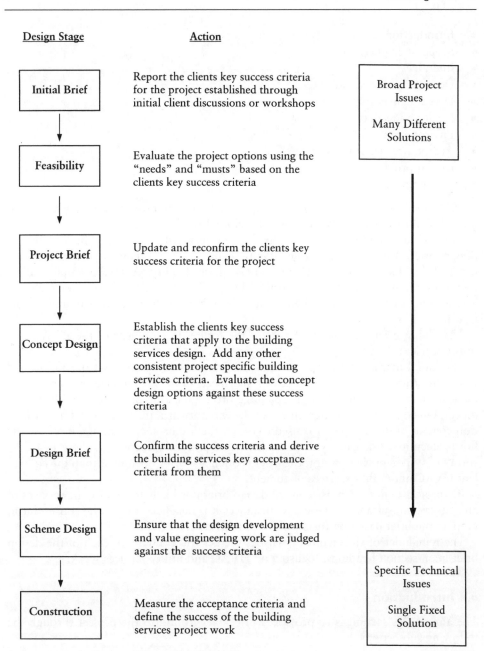

Figure 6.1 The design decision making audit trail.

6.3 The contents of a typical design brief document

The format and contents of a design brief will vary to suit the project requirements, however, the function will remain the same. For guidance, the section headings of a typical building services design brief are:

- Introduction
- Scope of engineering
- Interfaces
- Design standards and criteria
- Design deliverables
- Health and safety
- Environment
- Acceptance criteria
- Design programme
- Procurement

Other sections may be required to explain any unique project issues such as technical design risk or restraints imposed by phased working, etc.

The size and content of the individual project will determine the number of design brief documents. A large, complex project may require several separate design brief documents for different areas of the facility. A small, simple project may only require a single design brief. (The use of design brief documents should also extend into the other engineering disciplines such as electrical and public health services.)

The design brief sets out the basis for the future development of the HVAC and utility services.

The initial mechanical fire protection requirements are usually set out in a broad based fire strategy document that covers all aspects of the fire engineering. This document forms the basis for the concept stage fire engineering discussions with the fire officer. On completion and agreement of the way forward, the mechanical fire protection content of the document is used to create the fire protection design brief. As the fire protection work usually requires little design development, e.g. there is a LPC sprinkler system or there is not, the brief is often a shorter document than the equivalent HVAC and utility services document.

If a project requires a noise control design brief it is likely to require the services of an acoustic consultant. In these circumstances it is beneficial to ask the consultant to prepare the brief based on the section headings described earlier.

The remainder of the chapter describes the generic contents of each of the design brief main section headings, focusing on HVAC and utility services work.

6.4 Introduction

The aim of this section is to provide a brief background to the project through the following subsections.

6.4.1 General

A couple of short paragraphs describing the engineering discipline, the client, the site address and who is responsible for this section of the works within the design organisation, e.g.:

This design brief provides the basis for the design and specification of the…(type of system)…for…(client name)…at…(site address).

The…(type of systems, e.g. HVAC)…aspects of this project, as defined below will be the responsibility of…(name of designer and design organisation)

6.4.2 Project description and objectives

General description of the project works, ideally not more than a page and cross-referencing with the project brief, concept design report or other sources of pre-design information that apply to the project.

The description should explicitly list the success criteria used in the concept design evaluation matrix to ensure they are understood and integrated into the rest of the document, e.g.:'The current…(type of systems, e.g. HVAC)…'project requirements were established in the concept design decision analysis workshop held on 24 January 1999. Those requirements directly influence the building services design solution and are:

- Capital cost (add limit if known, e.g. £2 million)
- Revenue cost (add limit if known, e.g. £250,000)
- Programme duration (add limit if known, e.g. completion for February 2001)
- Reliability and maintainability in use
- Innovative design solution
- Maximum floor area
- Energy efficiency (add energy target if known)
- Provision for future expansion
- Speed of construction
- External plant requirements
- Zone control

6.5 Scope of engineering

6.5.1 Scope of services

The scope of services lists the complete range of services provided by the design organisation. The list should be based on the original conditions of agreement and include all services that will be provided from this point in the design process to completion of the agreement. Although the client may be aware of these activities it is unlikely that the rest of the team will fully understand the role. This is particularly important if your agreement is limited while other members of the project team have more comprehensive duties.

The list should represent the primary sections of work and will typically start with a review of the current project information and finish with witnessing of the commissioning and confirmation that the acceptance criteria have been achieved.

A scope of services list may include the following activities:

- Review the current project documentation, architectural and structural drawings

- Prepare design calculations; heat gain, heat loss, water storage, domestic water services demand loads, etc.
- Prepare system load assessments; heating and cooling loads, electrical loads, etc.
- Provide a thermal analysis simulation report for the central atria zone
- Attend project progress meetings at the architects' offices
- Liase with client staff to determine the services requirements
- Liase with local gas and water authorities
- Liase with building control (and the fire officer)
- Liase with the specialist gases, controls and containment booth subcontractors
- Communicate and co-ordinate the building services design activities with the other members of the project team
- Provide input to CDM design safety reviews
- Provide input to constructability reviews
- Prepare a scheme design ±15 per cent cost of work estimate
- Update the design brief document to reflect any basis of design changes
- Contribute to the value engineering workshops
- Provide schematics for the main systems: LTHW, utilities, ventilation, etc.
- Provide pipework and ductwork general layout drawings
- Provide detailed layouts for plantroom areas
- Provide equipment selections: AHUs, chillers, boilers, pumps, grilles, etc.
- Provide a mechanical services materials specification
- Provide a mechanical services detailed design specification
- Provide mechanical services performance specifications for the specialist gases, compressed air, containment booth subcontractors and asbestos strip-out contract
- Contribute to the project programming
- Prepare a detail design ±10% cost of work estimate
- Discuss with the quantity surveyor the tender short list and advise on the appropriate procurement routes
- Liase with the quantity surveyor to ensure the contract documentation is consistent with the technical documentation
- Attend the contractor tender interviews and assist in the contractor selection
- Review the contractors construction design information
- Advise on building services design variations
- Advise the quantity surveyor on interim valuations and payments to the contractor
- Attend site once a fortnight to review site progress and attend the construction review meetings
- Witness the building services systems testing and commissioning
- Confirm the acceptance criteria measurement results
- Ensure comprehensive operation and maintenance documents are issued to the client.
- Assist the quantity surveyor with the building services work final account

6.5.2 Scope of work

The scope of work lists the main building services systems developed by the design organisation. The list may be incomplete or only available as a broad set of systems at this stage in the design process. The aim is to record the current understanding of the building services works and provide a start reference. The document will continue to be revised throughout the project either as more information becomes available or as design variations occur.

A scope of work list may include the following activities:

- Removal of all asbestos products
- Isolation and strip-out of all the existing services within the project boundary
- Potable water distribution system from the tie in connection (located within 1 m of the site boundary supplied by the utility provider).
- Local point of use hot water services distribution.
- Natural gas distribution system from the tie-in location (located at the site boundary and supplied by the utility provider) to the boiler equipment
- 2 off, 60% load compressors and compressed air distribution system
- 2 off, 60% load, roof mounted air cooled liquid chillers and chilled water distribution system for each of the cooling zones
- 3 off, 40% load natural gas boilers and LTHW distribution system; condensing lead boiler
- Specialist gases (nitrogen, hydrogen and acetylene) bottle store, distribution network and detection system
- BS 5295:1989, 'Unmanned' Class K air conditioning system to the production area
- Twenty-four hour mechanical ventilation to the pressurised lobbies segregating the zone 2 hazardous areas from the unclassified areas
- Dust extract ventilation from the solids handling bag slit stations
- Solvent extract ventilation from the liquid handling equipment
- Natural ventilation to the central atria zone
- Mechanical ventilation and frost protection (electric) to the mechanical services plantrooms
- Mechanical services control system (BMS) with PC access located in the maintenance room
- Services metering to the water and natural gas supplies

6.6 Interfaces

In any project the number of interfaces are considerable. The aim of this section is to ensure that the whole range of interfaces, people and systems are understood.

6.6.1 Team interfaces

This section typically lists the client, the architect, the structural engineer, the quantity surveyor, the utility suppliers, the construction manager, the fire officer, the clients' insurer and any specified preferred suppliers, e.g. controls or containment booth suppliers. The list provides the name and contact address, telephone number, etc. and the reason for the interface.

6.6.2 Services interfaces

The services interfaces define all the system relationships with the other engineering disciplines. The physical interfaces need to be explicitly detailed, e.g. the electrical section will supply and connect a 415 volt power supply into the building services control panel. The interface definitions include utility services, public health, builders' work, architectural work and tie-ins to existing client or utility services systems.

6.7 Design standards and criteria

During the concept design work the review of the sources of design information will have highlighted the main standards and criteria that apply to the project work. This section of the design brief lists this information and confirms that this is the basis for future design development.

The section should include the following subsections:

6.7.1 Design codes and standards

This section lists all the relevant design codes and standards that are going to be applied to the design work. These vary with each project, although reference to Section 5.4 will provide a useful starting point. The list is often considerable and it is frequently beneficial to break the list down into subsections, e.g. general, pipework systems, ductwork ventilation systems, automatic controls, thermal insulation, etc.

6.7.2 Design criteria

Figure 5.2 and the concept design work will have established the building services design criteria for the project. Often the most convenient method of recording the internal space criteria is in a table, listing the different spaces and the primary environmental requirements and assumptions. A blank sample sheet is presented in Figure 6.2.

The external winter design criteria will need to consider:

- Thermal overload and preheat capacity
- Twenty-four hour operation and the effect of low night time temperatures on heater battery coil requirements

The selection of the external summer design criteria needs to carefully consider:

- The tendency for summer peak temperatures to rise above previous norms
- The cost effect on air cooled chillers and similar plant of selecting high design summer temperatures

6.8 Design deliverables

The design deliverables are the method of communicating the technical design proposals to the project team and specifically the building services contractor. Design deliverables generally consist of drawings, specifications and equipment schedules, though

Room	Design Temp. (+/- 2°C)	Design RH % (+/-10%)	Vent Ac/Hr (min)	Diff. Press (Pa)	Highest Filtration Level	Room Class. (BS 5295)	Number of Occupants	Equip heat gain (W)	Hazardous Area	Noise level (NR)	Comments

Figure 6.2 Sample internal design criteria sheet.

sometimes design calculations, wind studies, mathematical modelling, value engineering reports, cost estimates, etc. form part of the deliverables package. The form and content of the deliverable is driven by the approach to procuring the new works, e.g. 'performance' deliverables will be significantly different from 'detail design' deliverables'.

The description of the design deliverables needs to confirm the type and version of software used in producing the work. The deliverables are usually described as a list of drawings, specifications and equipment schedules. An example of this is:

6.8.1 Typical design deliverables

All drawings will be supplied in AutoCAD 14 format. All documents will be produced using Microsoft Word or Excel.

Reports

- Thermal analysis simulation report
- Scheme design report
- Feedback from the value engineering workshops

Schematic drawings

- BS 5295:1989, 'Unmanned' Class K clean area ventilation schematic
- Dust and solvent extract ventilation schematic
- Atria zone natural ventilation schematic
- Hazardous area lobby pressurisation schematic
- Specialist gases schematic
- Compressed air schematic
- Domestic water services and natural gas schematic
- Chilled water schematic
- LTHW and boiler plant schematic

Layout drawings

- BS 5295:1989, 'Unmanned' Class K clean area ventilation layout drawings (3 no @ 1:50 scale)
- Dust extract, solvent extract and lobby pressurisation ventilation layout drawings (2 no @ 1:50 scale)
- Atria zone ventilation layout drawings
- Specialist gases, compressed air, domestic water services and natural gas pipework layout drawings (4 no @ 1:100 scale)
- LTHW and chilled water pipework layouts (6 no @ 1:100)
- General arrangement layouts of mechanical services plantroom (3 no @ 1:20)

Specifications

- HVAC services detailed design specification
- Mechanical services installation specification
- Specialist gases and compressed air performance specification
- Solids and liquids containment booths performance specification
- Asbestos removal performance specification

Cost of work estimate

- Scheme design ±15 per cent cost of work estimate
- Detail design ±10 per cent cost of work estimate

Note: a general description of the design deliverables will have been previously listed within the scope of services section.

6.9 Health and safety

This section outlines the design health and safety hazards. The results of the concept stage CDM design review provides the main source of information. Where hazards cannot be eliminated, the hazard and the proposed solution need to be recorded in this section of the brief. Typical examples would be:

1 Systems involving 'relevant fluids', as defined in the pressure systems and transportable gas containers regulations will designed in accordance with:
    a COP 37, Safety of Pressure Systems. Approved Code of Practice, 1990
    b HS(G) 39 Compressed Air Safety, 1990
    c HTM 2022 Medical Gas Pipeline Systems
    The written Scheme of Examination will be prepared by an independent third party employed directly by the client.
2 The design of the domestic water services will comply with the following publications:
    a L8, The Prevention or Control of Legionellosis (including Legionnaires Disease) Approved Code of Practice, 1991
    b HS(G) 70 The Control of Legionellosis including Legionnaires Disease, 1993
    c CIBSE TM 13 Minimising the Risk of Legionnaires Disease.
3 All fire dampers will be fully addressable and will be instructed to close on receipt of a fire alarm signal at the mechanical services control panel.
4 All exposed surfaces with the exception of radiators will be kept below 55°C. Radiator systems will only be used in office areas. These areas are only occupied by adult staff and the potential risk is reduced during most of the heating season by weather compensating the system flow temperature.
5 CE marking of the installation will be the responsibility of the building services contractor. Individual equipment components will be supplied with certificates of incorporation for inclusion in the technical file. The building services contractor will be contracted to use an independent third party to compile comprehensive operation and maintenance documentation.
6 Ventilation of internal spaces containing refrigeration pipework will be consistent

with the requirements of BS 4434:1995 Specification for Safety and Environmental Aspects in Design, Construction and Installation of Refrigerating Appliances and Systems.

6.10 Environment

This section describes the environmental issues associated with the proposed design. Typical examples would be:

- The primary fuel source will be a natural gas (CO_2 equivalent of 0.2 kg/kWh)
- The base load boiler will be a condensing boiler and the LTHW system design will include weather compensation directly on the boiler to maximise condensing
- Perimeter rooms will have natural ventilation with high level and low level windows
- The central atria zone will have natural ventilation for summer time cooling
- Building design orientation (in conjunction with the architect) will be optimised to minimise annual energy consumption, maximise daylighting and reduce solar gain
- The building envelope (in conjunction with the architect) will be designed, constructed and tested to achieve a best practise maximum air leakage as described in BSRIA Air Tightness Specification 10/98
- The design of the ventilation systems will include free cooling and recirculation of tempered and conditioned air where this is consistent with the safe and effective operation of the facility
- The domestic water services design will incorporate low flush cisterns, showers (rather than baths) and spray taps; urinal systems will be automated to flush after use
- The package chiller selection will be based on non-ozone depleting R134A refrigerant; there will be no 'site run' refrigerant pipework
- The discharge of dust extract systems to atmosphere will be filtered through a cyclone collector and double HEPA filer arrangement
- Solvent laden extracts will be incinerated prior to discharge to the atmosphere
- The complete building services installation will be controlled through a building management system (BMS); the system will provide the following control functions:

  - Weather compensation
  - System optimisation
  - Variable speed control on low load demand
  - Control of individual space conditions
  - Varying set point conditions in winter (20°C) and summer (24°C)
  - Twenty-four hour, 365 day programming

An annual energy target will be established during the scheme design stage.

6.11 Key acceptance criteria

The key acceptance criteria are measures that represent the objective definition of the success criteria and complete the design decision making audit trail. They can be any 'measurable' criteria that are 'controlled' by a building services system or involve the provision of the building services design and construction work. The obvious criteria are space temperature, humidity, noise rating, cost and programme. Less obvious criteria may be annual energy consumption, cleanroom particle count in manned operations, recorded OEL in operation, cleanroom clean-up rate, acceptable environmental noise level at the reference point, design within tender cost, installation sign-off by the fire officer, construction within the shutdown period, achieved design CHP plant output, no tenant complaints, etc.

On large or more complex projects it is likely that the building services work will have a number of key acceptance criteria. However, the chosen measurable values need to be reasonable (measurable criteria and a specific number) and appropriate within the limits of the design work (i.e. the criteria is controlled by the new building services system and the measurement time period is either within the commissioning period or a reasonable post-construction period). The acceptance criteria **must** be clearly specified to include:

- The measured value
- The acceptable variation in the measured value
- The precise location of the measurement position
- The acceptable operating conditions when the measurements are recorded
- How the values will be measured

Examples of acceptance criteria are:

1   $20 \pm 2°C$ at 1.5 m above finished floor level in the managing director's office. Continually measured between 09:00 and 17:00 h over three continuous days during the commissioning period in an unmanned condition. Measurements will be recorded by the new BMS system and paper copies of the daily results entered in the commissioning manual.

2   Class K (ref. BS 5925, 1989) particle count within 1 m of the product fill-pointed in four different horizontal locations. Measurements taken once in an 'As Built' (ref. BS 5925, 1989) condition during the post-commissioning and pre-handover period of the project. Record measurements will be made at all the BS prescribed particle categorises with a portable light scattering particle counter.

3   Tendered cost of work, £2.5 million $\pm$ 5 per cent, measured as contract award value for the proposed works. (*Note*: although the final account cost of work is probably the clients main concern, it is often heavily influenced by non-design issues, e.g. claims for delay due to lack of access or programme delays and consequently may not be an appropriate acceptance criteria).

4   The current design requirement provides an additional 300 kW boiler capacity and two valved connections on the balance pipe for future use. The additional capacity will be measured as a recorded increase in the theoretical design calculations.

6.12 Design programme

It is unlikely that a detailed programme of the future project works is available when the design brief is written. Consequently the aim of this section will be to outline the design programme intent using the best currently available information.

The focus of the section is the design part of the project programme although the programme should allocate reason periods to procurement, installation and commissioning (acceptance testing).

The design programme should define the separate design deliverables for each work package, e.g. design brief, reports, drawings, specifications, cost of work estimate, etc. It should also include the interim design milestones, e.g. design brief sign-off, value engineering review, scheme design review, scheme design sign-off, completion of the architectural and structural layouts, building control (fire officer) design acceptance, detail design review, tender issue, contractor interviews, etc.

6.13 Procurement

The design programme and method of construction procurement are closely interwoven. The method of procurement will have a significant influence on the design deliverables and construction management. A large number of small work packages, e.g. ductwork, air handling units, controls, smoke ventilation, etc. will require more design time and on-site co-ordination than a complete building services package placed with a single contractor. A single building services design package may require less design time although it may not be consistent with the philosophy of the management contractor (see Section 2.5.1).

The procurement section should define:

- The form of contract for the building services works; where the contractor is involved in design work it is essential that the form of contract reflects these duties
- Who is procuring the installation works, e.g. the client, the main contractor, the management contractor
- The proposed number and type of tender bid packages
- The number of bid invitations for each of the packages (including names of companies where applicable)

6.14 Design brief development

When the design brief is complete it needs to be issued to the project team for review, incorporation of comments and sign-off by the client. It will remain a 'live' document at least until the end of the design development period (end of scheme design). During the development period the design brief may need to be revised to incorporate the latest design developments. In these circumstances the original design brief should be revised, re-issued and signed-off by the client so that it remains up to date.

The design brief is the reference point for all the future design work and an essential tool in the design methodology.

6.15 Chapter review

This chapter describes the function of the design brief and explains how the function is achieved by outlining the generic contents of each main section of a typical design brief. The main sections of a typical design brief are:

1 Introduction
2 Scope of engineering
3 Interfaces
4 Design standards and criteria
5 Design deliverables
6 Health and safety
7 Environment
8 Acceptance criteria
9 Design programme
10 Procurement

Completion and sign-off of the design brief document is followed by the scheme design stage.

Chapter 7

Scheme design stage

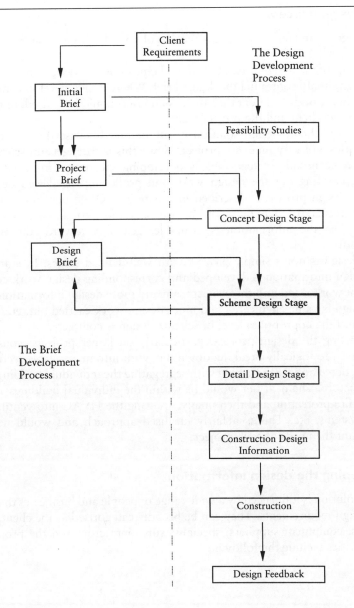

7.1 Chapter contents

The topics covered in this chapter include:

- Developing the design information
- Design work
- Value engineering
- CDM requirements
- Scheme design deliverables

7.2 The design process

Scheme design is an iterative process that results in the broad clarification of the complete design solution.

The contents of the design brief guide the scheme design development. No aspect of scheme design should contradict the design brief. Where design development results in a change from the original intent then the design brief should be revised, re-issued and signed-off by the client and the project team.

Almost every scheme design solution is unique and likewise the specific scheme design activities will vary on every project. While this is true, general scheme design activities can be broadly categorised as developing the design information, value engineering, carrying out the design work and producing the design deliverables. The scheme design process is described in Figure 7.1. Often developing the design information seems to take most of the time as the design work can only be completed when the information is available, generally towards the end of the design period!

Scheme design is not a simple process with well-defined stages. It is an iterative process, design information is developed and corresponding design work completed. The resultant work is fed into the project team and more design information is identified, the design is value engineered and more design work carried out, etc. The loop continues until the appropriate level of scheme design is complete.

Before starting the design process, particularly on larger projects, consideration should be given to logically breaking down the work into manageable sections. This can follow a tiered approach starting with segregating the individual buildings and the site infrastructure. The next tier would be within the individual buildings separating HVAC and fire protection and then finally, splitting the HVAC into ventilation and pipework systems, etc. This is a fairly idealised approach and would need to be adapted to suit the needs of each project.

7.3 Developing the design information

The design solution is influenced by a wide range of people and businesses that all have differing design requirements. They can be broadly categorised as the client, external third parties, equipment suppliers, specialist sub-contractors and the project team. Typical activities include the following.

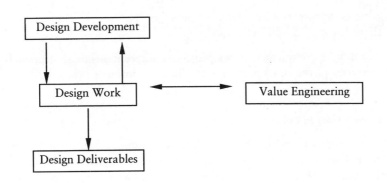

Figure 7.1 Scheme design flow diagram.

7.3.1 The client

Although the broad concepts have been established in the design brief the detail behind the concepts needs to be determined. Typically the following topics are discussed:

- Anticipated future demand requirements
- Existing and proposed, process and fabric loads
- Age, detail and capacity of the existing systems
- COSHH details of proposed systems and substances
- Confirmation of the proposed environmental design criteria
- Sensitivity of the client's staff to issues that affect building services
- Client preferred equipment suppliers
- Clients' client requirements

The discussions start where the concept design and design brief ended and it is quite likely that some information will be repeated from earlier activities.

7.3.2 External third parties

A number of external third parties influence the design. Typically these are people who are employed to ensure that the legislation and industry standards are correctly implemented or advisers to the project financiers who need to ensure that the design solution meets the financiers needs. Their requirements are often judged as mandatory and in many respects they are, although most are open to constructive discussion on suitable alternative design solutions. The types of contributors that affect the design outcome are:

- Building control (and through them the fire officer)
- Clients insurance broker and lead insurer
- Health and safety executive
- Environmental health officer
- FDA and MCA inspectors
- Statutory service providers, e.g. gas, water, electricity, telecommunication
- The clients' inspectors

- Consultants representing the project financiers
- The clients proposed and existing tenants

The list cannot be complete as every project is different although it probably represents the majority of the third party groups likely to influence the design.

7.3.3 Equipment suppliers

The primary equipment suppliers, e.g. chillers, fume cabinets, condensing boilers, pumps, etc. will all have their own unique equipment requirements that need to be integrated into the scheme design. Equally they will all have their impact on the design solution.

The types of issues that can occur are multifarious and depend on the specific project and equipment requirements. An indicative sample for a range of equipment would be: space and access requirements, noise emissions, chiller switching frequency, storage capacity, minimum pump suction head, filter efficiency, minimum flow rate and temperature, absorbed power, maximum pressure, hazardous rating implications, GMP requirements, storage life, availability, controls interface requirements, environmental conditions, etc. All these types of issues need to be determined before design work can be completed. In some instances the design solution will be based around a specific equipment suppliers system.

7.3.4 Specialist subcontractors

In certain instances the scheme design work is carried out by specialist subcontractors as the design organisation may not have the necessary skills, qualifications or time to successfully complete the work. The type of specialist contractor will depend on the project requirements, typically this could include:

- Fire protection systems, most building services design organisations are not LPC approved fire protection designers
- Building services control systems, e.g. Trend, Johnson Controls and other controls contractors
- Acoustic consultants
- Specialist warehouse ventilation contractors
- Process containment systems contractors

The specialist subcontractors' requirements will need to be considered in the same manner as the equipment suppliers and the system implications integrated into the scheme design.

7.3.5 The project team

The building services scheme design cannot develop in isolation from the rest of the project team, any more than the other disciplines can act in isolation. Consequently the design information needs to be fed to the other members of the project team and integrated or adapted (in consultation) into the whole project design.

The architectural discipline invariably has the largest impact on the building services systems. It is essential that a 'two-way' flow of information is established early in the design process. This should ensure that the building services space requirements are accommodated and that the systems are integrated into the building fabric.

The design brief outlines the interfaces between the project team. These need to be effectively managed throughout the scheme design process.

7.3.6 General

When the scheme design stage is complete all the main design information should be known and agreed with the project team. The type of system, its spatial requirements, the primary items of plant, the indicative size of the plantrooms, the material selections, etc. should all have been finalised.

Detail design will consider items such as dimensioned routing of ductwork in the building, co-ordination, selection of valves and dampers, etc. This level of information is generally unnecessary and premature for scheme design.

7.4 Design work

The extent of the design work is determined by the type and consequently, purpose of the scheme design deliverables. These issues will already have been addressed in the design brief. The type of design deliverables are defined in Section 6.8 and the purpose in Section 6.13 (i.e. the purpose is defined by the how the work is going to be procured). Frequently the options reduce to either performance deliverables for design by the contractor or detail design deliverables for installation by the contractor

The type of design work required at this design stage is determined by the nature of the project. It could include:

- Heat loss and heat gain calculations
- Fluid flow modelling or analysis thermal simulation
- Wind tunnel testing
- Air conditioning calculations
- HVAC ductwork and pipework calculations
- Plant and equipment selections including factory review of sample products, where appropriate
- Mass loading estimates for the structural engineer
- Electrical load estimates for the electrical engineer
- Environmental noise calculations
- Fire water demand and underground main pipework calculations
- Utility services loads and pipework calculations
- Layout and integration of the HVAC and fire protection systems into the building
- Confirmation of the preliminary HVAC plantroom locations, sizes and access requirements
- Preliminary system schematics and layout sketches
- A HVAC and utility services technical performance specification or scheme design report for the proposed systems
- A fire protection technical performance specification

Pharmaceutical projects frequently require design calculations to be submitted as design deliverables. In these instances it is essential that the prescribed format for the calculations is understood and applied. (These requirements will have been described in the design brief document.)

7.4.1 Calculations

All calculations need to be prepared according to the appropriate quality assurance procedures. They need to be quickly and easily understood as they invariably need to be revised and will need auditing. It is also worth remembering that contract liabilities can extend up to 10 years after project completion and calculations may need to be retrieved as part of a future claim. Consequently they must be clearly structured and held as a hard copy in the project calculation file and where appropriate electronically in the project file. Complete calculations should include:

- A date, a unique project reference, sequential page numbering, the name of the person who prepared it and the person who checked it
- A section describing the calculation purpose
- A section listing all the reference sources used in the calculation
- A clearly structured calculation so that anybody could pick it up and follow it through
- A conclusion, e.g. the boiler selection will be based on a total design load of 800 kW

7.5 Value engineering

Value engineering is an opportunity to 'fine tune' the best fit solution and remove any unnecessary cost. It represents the third stage in the evaluation process that started with project evaluation in the feasibility stage and follows design evaluation in the concept design stage [1].

A value engineering workshop should be organised when the design work has developed schematics and layouts for the main systems and the form of the final design is reasonably well known. The workshop should include the key people involved in the building services solution, e.g. the client's project manager, the quantity surveyor, the architect, the contractor, etc. (Consider involving a contractor from a previous project if the project contractor has not yet been appointed.) The process follows the stages described in Figure 7.2.

The key cost items are the systems or components that offer the greatest opportunity for cost saving. When these are identified their function in the design solution needs to be established and recorded. Functional analysis system technique or FAST diagrams are recommended [1] for this purpose. These break down the primary function of the system or component into a series of subsystems that describe how the function is achieved.

The aim is to establish a common understanding of why the system or component is required. The next step is to brainstorm as many alternative design solutions as possible that achieve the same function (assuming all the original functions are necessary). The alternatives that appear to provide a better value solution for the same function-

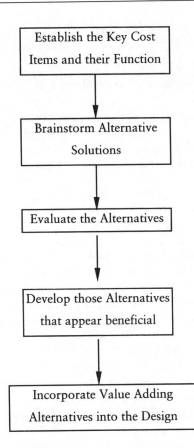

Figure 7.2 Value engineering flow diagram.

ality are established by the workshop. Where possible the alternatives are developed within the workshop and those that have a proven benefit are introduced into the design solution. In many instances the development of the alternatives needs to be progressed outside the workshop. These are then brought back to a future workshop or close out meeting to confirm if they are going to be introduced into the design solution.

7.6 CDM requirements

The CDM design requirements in the scheme design stage are essentially the same as concept design requirements, i.e.:

- Identify the significant health and safety hazards and risks of any design work
- Give adequate regard to the hierarchy of risk control

However, the level of design will now be significantly more detailed and new health and safety risks may have been introduced to the design.

Following on from the process described in Section 5.9, and the health and safety section in the design brief. The scheme design health and safety review should cover the following:

1   Re-confirm the building services scope of work and the physical boundary limits of the proposed work.
2   Review the check list of topics detailed in the health and safety hazard check list (Section 5.9).
3   Review the project list of hazards and the proposed actions to reduce or eliminate them. These will tend to be broad issues with general responses, e.g. all systems involving 'relevant fluids', as defined in the Pressure Systems and Transportable Gas Containers Regulations will be designed in accordance with:
    • COP 37, Safety of Pressure Systems. Approved Code of Practice, 1990
    • HS(G) 39 Compressed Air Safety, 1990
    • HTM 2022 Medical Gas Pipeline Systems.
4   Confirm these issues have been actioned.
5   Establish where these general hazards or any others introduced during scheme design now present specific hazards. Specific hazards are the type of hazards you may recognise in a normal working day, e.g. opportunities for slips, trips, or falls, propped open fire doors, hot surfaces, manual handling issues, mud on roads, safety valve discharges not piped to a safe location, valve access requiring a ladder, pipes located at head height, etc.
6   List all the specific hazards.
7   Highlight the remaining hazards that represent a 'significant' risk (see below).
8   Apply the hierarchy of risk control (see Figure 5.3).
9   Record how the project risks are going to be controlled and ensure these actions are integrated into the design solution.

After completion of the CDM review any 'significant' hazards inherent in the final design solution must be recorded and passed to the planning supervisor.

7.7 Risk assessments

Risk assessments are a method of converting a subjective hazard into an objective answer that allows the risk to be controlled. The following definitions are taken from Ref. [2]:

Hazard   The potential to cause harm.
Risk   The likelihood that harm will occur. The size of a risk is determined by the *likely severity of harm* and the *likelihood that harm will occur.*

A simple risk assessment would follow the following sequence of actions:

1   Identify the hazards (see previous section)
2   Categorise the severity of harm:
    a   Very low, e.g. a scratch or bruise
    b   Low, e.g. a cut that might require a stitch

Table 7.1 Risk evaluation chart

Severity of harm	Likelihood of harm occurring				
	Not likely	Possible	Quite possible	Likely	Very likely
Very low	Very low	Very low	Low	Medium	Medium
Low	Very low	Very low	Medium	Medium	High
Medium	Low	Medium	Medium	High	High
High	Medium	Medium	High	Very high	Very high
Very high	Medium	High	High	Very high	Very high

    c   Medium, e.g. loss of a finger tip, a broken leg
    d   High, e.g. loss of an arm
    e   Very high, e.g. a fatality
3   Categorise the likelihood that the harm will occur (if there are control measures in place, categorise the likelihood with them in operation):
    a   Not likely, e.g. very infrequent, like a lightning strike
    b   Possible, e.g. will require a number of factors to be present if it is going to occur
    c   Quite possible, e.g. a fault would be necessary to cause the event
    d   Likely, e.g. an accident may occur though infrequently
    e   Very likely, e.g. an accident will occur
4   From Table 7.1 evaluate the risk

Any risk greater than a medium can be considered 'significant' and consequently requires application of the hierarchy of risk control.

A sample chart for recording project hazards and how they are going to be controlled is presented in Figure 7.3.

7.8 Scheme design deliverables

7.8.1 General

The content of the scheme design deliverables is determined by the procurement route.

Traditional and construction management contracts usually require a level of design detail that allows the work to be procured on an installation contract with a minimum of design work by the contractor. These types of contract require the design to be developed to the end of the detail design stage (see Chapter 8).

The scheme design deliverables on this procurement route will probably be a short technical design report and a set of scheme design drawings for a project team review and sign-off. The review would address the following type of issues:

- Establish progress against the design programme
- Clarify any remaining technical issues
- Ensure the design development is consistent with the design brief and client requirements
- Confirm the design has been integrated into the overall project work

Risk Assessment and Control Record Sheet

Project Number :				**Reference Information :**		**Page Number :**
Hazard Location :						**Assessor :**
Type of Hazard :						**Assessment Date :**
Hazard Description	Severity	Likelihood	Risk	Control Actions		Comments

Figure 7.3 Risk assessment and control sheet.

Design and build contracts place a significant design requirement on the contractor in addition to the installation work. In these circumstances the level of design detail required for contract issue is less than the traditional route as the contractor only needs to be informed of the design performance requirements and an outline of the proposals. However, the scheme design deliverables for this type of contract are more substantial as they are issued for tender at the end of the scheme stage. In these circumstances the main scheme design deliverables are performance design specifications and scheme design drawings.

The remainder of this chapter assumes that the scheme design deliverables are being produced for a design and build contract.

7.8.2 Starting the design deliverables

Design deliverables have a simple and difficult aim, to unambiguously communicate the design intent. In practice good design deliverables minimise ambiguity. Consequently before producing any design deliverables, the following issues should be considered against the specific project requirements

1   Design deliverables should only describe the technical requirements of the proposed works. Generally the quantity surveyor will provide the form of contract and contract preliminaries. However, the designer must review both these documents to ensure they are compatible with the design deliverables. The form of contract should detail the contractors' design responsibilities. The contract preliminaries should adequately describe the work's requirements. Where the documents are not adequate the deficiencies should be highlighted to the quantity surveyor and the documents revised. Issues such as providing schedules of rates, provision of site management, working with other trades, tender return breakdown, etc. may have to be incorporated into the technical specifications.

2   When the contract has been signed by the successful contractor the balance of 'control' over the work package has significantly shifted from the designer to the contractor. The only 'tools' that can be applied to control the work package are the contract documents (form of contract, contract preliminaries and design deliverables, i.e. performance specification, scheme design drawings, detail design specification, etc.). Generally errors or omissions in the technical specification and drawings will only be rectified by acknowledging the mistake (never very pleasant!) and asking the client for more money. This will probably jeopardise one of the acceptance criteria (achieving the cost target) and reduce the overall success of the project.

3   The level of detail in the design deliverables reflects their aim. Arguably, the higher the level of detail the less opportunity there is for the contractor to misinterpret the intent. However, this has drawbacks, particularly with performance deliverables as more detail restricts the opportunity for the contractor to add their experience and by implication suggests a design solution that the project should adopt. However, as the contractor provides the detail drawings, the scheme drawings only need to provide enough information to allow the contractor to do the detail design. Likewise, detail design deliverables only need to provide enough information to allow the contractor to produce the construction design information.

4   Ensure that quality assurance procedures are followed and that all design deliver-

ables are checked and audited before tender issue. Nobody has all the good ideas and frequently working on a project for a period of time does not allow the designer to 'see the wood for the trees'.

5   If there is a standard format for design deliverables use it.

6   Involve a contractor in a technical review of the design deliverables before they are issued for tender. If this is not possible organise an internal peer review of the documents. This provides an excellent opportunity to 'tighten-up' the deliverables before the contractor is signed up.

7   The conventional form of design deliverables has been paper based specifications, drawings, etc. Although these methods have proved robust it is always constructive to challenge there continued use. Currently they remain the preferred method of communication although it is likely they will be replaced with alternatives, designed to suit the available information technology, e.g. a CD-ROM with hypertext or electronic posting of documents in a central project document management system.

8   Ensure the full scope of work defined by the design deliverables is known before starting the work, e.g. where does the electrical engineering design package tie into the HVAC work package? Who is providing the sanitary equipment schedule? This should have been resolved in the design brief although occasionally new items are identified.

7.8.3 *Structure of specifications*

Most design organisations will have their own standard format for specifications. Frequently these are based on standard repeatable formats such as the National Engineering Specification (NES) [3], NHS Estates Model Engineering Specification (for health care work), or the old Property Services Agency (PSA) documentation, e.g. M&E 100. Alternatively they may use a bespoke combination of a project specific specification (describing the unique project works either in performance or detail terms) and an installation specification (describing the installation quality requirements for a range of building services systems).

Information describing specification requirements is provided in Refs. [4,5].

Where no repeatable specification exists and products such as NES are not used, an alternative format for a project specific specification would be:

- A cover page with project title and QA sign-off banner describing who prepared, checked and approved the document, the issue date and the revision number
- A page header with a project number and title, specification title, page number and revision number
- A page footer providing a document reference
- A contents page
- A revisions page
- A definition of the terms used in the document
- The technical specification (see sections below)
- Appendices: these contain any other reference information necessary to describe the works and explain the project requirements. Typically these include plant and equipment schedules, a list of the drawings describing the contract works, photo-

graphs of the site, the quality plan requirements (see Section 10.4) site operating requirements, a list of the preferred suppliers, the installation specification (see Section 7.8.5), a proposed programme for the contract work, a design responsibility matrix (see Section 7.9.2), an interface matrix, a CDM hazard list (see Section 8.7), a tender summary breakdown, etc.

7.8.4 HVAC and utility services technical performance specifications

Technical performance specifications provide specific descriptive requirements for the project work.

The section headings within a typical performance specification are usually the same as they prompt the type of detail that applies to all projects. They are:

- Scope
- Scheme design description
- Specific design requirements
- Project specific requirements
- Acceptance criteria

The remainder of this section describes the generic content of each of the main section headings.

7.8.4.1 Scope

The aim of this section is to provide a brief introduction to the project and describe the proposed works. The section comprises of the following subsections:

(a) *Introduction*: a short paragraph describing the type of work, the client and the site address, e.g. 'This performance specification provides the basis for the design and installation of the HVAC systems and equipment for...(client name)...located at...(and site address).'

(b) *Scope of services*: a list of the services that the contractor provides as part of the contract works, e.g.

- Detail design and installation of the building services scope of work
- Provision of design deliverables
- Provision of CDM assessments
- Provision of a quality plan
- Attend design team meetings
- Provision of the operation and maintenance manuals
- CE marking of the system

(c) *Scope of work*: a list of the main systems and all associated enabling work required within the contract works, e.g.

- A supply and extract air conditioning ventilation system to the dispensing suite
- LTHW pipework system to the air conditioning and fan coil heating system
- A BMS control system integrating all the proposed building services systems complete with remote PC monitoring and control equipment

- Commissioning of the all building services systems, etc.

7.8.4.2 *Scheme design description*

The aim of this section is to ensure that the contractor fully understands the HVAC system requirements. The section describes the scope of each of the work items listed in the previous section.

7.8.4.3 *Specific design requirements*

The aim of the section is to detail the basis for the design and outline the contractor's design deliverables. The section describes the criteria that the designer will use to monitor and control the contractors 'future' design work.

The individual project requirements determine the precise content of this section; however, as a minimum it contains the following subsections:

(a) *Codes and standards*: a list of all the design standards that the contractor shall apply to the design work. The list may be split into general standards that apply across all the project work, e.g. The Building Regulations and individual standards that apply to specific systems, e.g. BSRIA, The Commissioning of Air Systems in Buildings, Application Guide 3/89 1989. All the design standards listed in the design brief should be listed in this section.

(b) *Design criteria*: a table of all the internal and external design criteria that apply to the project. This should be an updated copy of the details in the design brief document.

(c) *Specific design requirements*: a list of all the specific technical design requirements that should be applied when carrying out detail design, e.g.

- Minimum external, fresh air supplied per person in occupied areas shall be 12 l/s; design fresh air requirements shall be taken from CIBSE Guide Table A1.5
- Ductwork velocities shall be maintained between 2.5 (final connections) and 7.5 (main duct runs) m/s
- Ductwork pressure drops shall be maintained between 0.6 and 1.2 Pa/m
- All fan systems shall be capable of maintaining 110 per cent of the design flow rate against 115 per cent of the design system pressure; this includes all cable sizes, motors, starters, etc. involved in the electrical power system
- Chilled water and LTHW pipework velocities shall be maintained between 0.75 and 1.5 m/s, etc.

(d) *Construction design information*: a list of all the design information (including the number of copies) the contractor needs to produce to ensure that the contract design requirements have been satisfied, e.g.

- Air conditioning psychometric calculations
- Ductwork layout installation drawings (1.50 scale)
- LTHW and chilled water system schematics (NTS)
- Plantroom layout installation drawings (1:20)
- Manufacturers plant and equipment drawings, including control panel and controls wiring drawings, etc.

7.8.4.4 Project specific requirements

This section is an opportunity to explain to the contractor any issues that may be considered unusual or unique on the project and that impact on the contract works but are not specific to HVAC. Typically this includes issues such as GMP, validation, food hygiene requirements, etc. This section describes the impact of this type of requirement on the building services contract works.

7.8.4.5 Acceptance criteria

The aim of this section is to explicitly state the acceptance criteria for the contract works.

The section will repeat selected key acceptance criteria agreed with the client in the design brief document and where necessary explicitly define other acceptance criteria, e.g. acceptable commissioning flow rate tolerances, design noise rating levels in the occupied space, etc. Any additional acceptance criteria need to be specified in the format described in the design brief section on key acceptance criteria.

7.8.5 Installation specification for HVAC and utility services work

The installation specification details the installation quality requirements for the systems described in the performance specifications. The broad nature of building services means that these types of documents are very large, as they endeavour to cover every installation aspect of every system. The master installation specification is reviewed by the designer to ensure that:

- The systems proposed in the contract works are comprehensively and correctly detailed
- The installation requirements proposed are in accordance with the specific project requirements
- The sections of the document that are not relevant are removed to avoid confusion

The content of master installation specifications are built up over a considerable number of years with the experience of each project fed back into the document. Generally they are the copyright property of the design organisation and are not repeated in this text however the contents typically include:

- Installation standards
- Pipework services
- Ductwork services
- Thermal insulation
- Testing and commissioning
- Operating and maintenance documentation

Note: this section is repeated in Section 8.9.2.

7.8.6 HVAC and utility scheme design drawings

The drawings supplement the specifications and describe the physical requirements of the systems and their integration within the building fabric. Precise guidance is not possible although the following points should be considered when preparing the drawings.

- Provide only as much detail as is necessary to communicate the design intent.
- Use schematic and diagrammatic drawings
- Pipework and ductwork layouts may be undimensioned single line drawings (1:100 scale not 1:50)
- Where plantroom layouts are required these may be issued as block diagrams rather than detailed drawings
- Provide standard design details, e.g. commissioning sets, coil details, sleeved penetrations through walls, etc.
- Only provide dimensioned information where a detail is required to confirm that the proposed design fits within the space envelope

7.8.7 Fire protection technical performance specifications

The fundamental difference between a fire protection and HVAC utility services performance specification is that the fire protection performance specification is based on a set of prescribed standards. Depending on the clients' corporate insurer these standards are generally LPC rules (UK standard) or NFPA codes (American standard). These standards describe the design and installation of fire protection systems. The fire protection specification is simply a guide that specifies the requirements that the prescribed standards omit and clarifies choices that are built into the standards.

The key to writing a successful fire protection specification is understanding the content and limitations of the prescribed standards.

The format of a fire protection performance specification is similar to the HVAC utility services performance specification except that it provides more installation and operation and maintenance information as the specifications are generally issued without the equivalent of a installation specification. The primary section headings are:

- Scope
- Scheme design description
- Specific design requirements
- Project specific requirements
- Supply and installation
- Acceptance criteria
- Operation and maintenance documentation

The remainder of this section describes the generic content of each of the main section headings.

7.8.7.1 Scope

The content of the scope section is consistent with the HVAC and utility services performance specification requirements except for the addition of a process description subsection. The process description is inserted after the introduction and before the scope of services and scope of work subsections. The fire protection contractor needs to understand the type of process, how it operates, the materials handled (particularly flammable and highly flammable liquids) and the materials storage so that the fire risk is clearly understood.

7.8.7.2 Scheme design description

The aim of this section is to ensure that the contractor fully understands the fire protection system requirements. This section describes each of the scope of work items listed in the previous section.

7.8.7.3 Specific design requirements

The aim the of section is to detail the basis for the design and outline the contractor's design deliverables. The individual project requirements determine the precise content of this section, however. as a minimum it contains the following subsections.

(a) *Codes and standards*: a list of all the design standards that the contractor shall apply to the design work, e.g. the LPC rules for automatic sprinkler installations including all current technical bulletins, NFPA 16, 1991 Deluge Foam-Water Sprinkler and Foam-Water Spray Systems, etc. All the design standards listed in the design brief should be listed in this section.

(b) *Specific design requirements*: a list of all the specific technical design requirements that would apply when carrying out detail design, e.g.

- The main production plant storage and remote warehouse shall be provided with high hazard category II sprinkler systems
- The minimum design density for these systems will be 12.5 l/min per m^2, compliant with type S1, high piled storage, not more than 5.9 m high
- The deluge valve header shall be provided with two current and three future use connections
- The water-foam deluge systems shall be designed to provide a minimum design density of 6.5 l/min per m^2 over the storage tanks
- The indicative flow range of the proportioner taking into account maximum future use is estimated to be 1,650–11,000 l/min
- Flow switches down stream of the control values shall be linked into the fire alarm and provide remote alarm indication in the control room

(c) *Construction design information*: a list of all the design information (including the number of copies) the contractor needs to produce to ensure the contract design requirements have been satisfied, e.g.

- A certificate of compliance that confirms the contract works comply with the design and installation requirements
- Sprinkler hydraulic calculations and associated schematics
- Sprinkler pipework working drawings (1:50 scale)
- Deluge head details

- Compressed air and charging rate details
- Synthetic foam concentrate details

7.8.7.4 Project specific requirements

This section is an opportunity to explain to the contractor any issues that may be considered unusual or unique on the project and that impact on the contract works but are not specific to fire protection. Typically this includes issues such as GMP, validation, food hygiene requirements, etc.

7.8.7.5 Supply and installation

Most aspects of the basic supply and installation requirements are detailed in the prescribed standards. Occasionally the prescribed standards do not adequately describe the project requirements. In these situations the additional requirements have to be explicitly stated. Examples of this are:

- Pre-installation and post-installation painting of sprinkler pipework
- Not allowing G clamp LPC approved pipework fixings, if appropriate
- Use of MDPE fire water and hydrant underground pipework and fittings
- Labelling of pipework and valve sets
- Earth bonding provision to BS 7671 IEE regulations of electrical installations
- Pressure testing of dry riser installations
- Provision of spares, particularly the supply of one complete charge of foam for testing purposes

7.8.7.6 Operation and maintenance documentation

The operation and maintenance documentation in the prescribed standards frequently does not adequately define the project requirements. In these situations the additional requirements have to be explicitly stated. Examples of this are:

- Presentation and content of the O&M manuals
- AutoCAD drawing files for the contract works
- Supplier LPC approval certificate
- Hydraulic calculations for each system
- Pipework pressure test certificates

7.8.7.7 Acceptance criteria

The aim of this section is to explicitly state the acceptance criteria for the contract works.

The section will repeat selected key acceptance criteria agreed with the client in the design brief document and where necessary explicitly define other acceptance criteria, e.g. multiple hydrant flow rate and pressure testing requirements, multiple hose reel flow rate and pressure testing requirements, time taken to achieve the design flow rate over a deluge system. Any additional acceptance criteria need to be specified in the format described in the design brief section on key acceptance criteria.

7.8.8 Fire protection scheme drawings

Fire protection scheme drawings do not require the same level of consideration as the HVAC and utility services drawings. The following types of drawings usually provide adequate detailing of the scope of work.

- Project general arrangement drawings with the area requiring fire protection coverage shaded; these drawings need to show the plan dimensions, height of the area and voids within the construction
- Drawings of the process pipework, racking, etc. and equipment in the areas provided with fire protection; in the case of deluge systems the details of the equipment being protected should be included
- Single line undimensioned underground pipework layout drawings
- Schematics of the proposed fire protection systems

7.8.9 Noise control requirements

The nature of noise control is not discipline specific and there is no physical noise control system to procure. Noise control requirements, i.e. enclosures, low noise valves, attenuators, structural mass, sound reducing glazing, etc. need to be included in the different engineering technical specifications.

The noise consultant will create a noise 'software simulation model' of the project, advise on the overall project design and provide detailed noise control specifications for all the plant and equipment procured by the individual discipline designers. If the environmental noise levels have not yet been established the consultant would record them at the agreed reference points, prior to any work starting on site and ensure the EHO accepted their data.

During the contract works the consultant would review all the manufacture's plant and equipment noise data against their contract specifications and resolve any associated design issues. On completion of the work their role would involve witnessing commissioning tests, troubleshooting problems and re-measurement of the environmental noise levels at the prescribed reference points to ensure the EHO's criteria had been met.

7.9 Scheme design responsibility

7.9.1 General

Establishing the split in design responsibility between the designer and the contractor on design and build projects is superficially easy although in reality it often proves to be quite difficult. Although the design organisation may prefer to place all the design responsibility on the contractor in all except the simplest cases this is often not practical. Only projects that involve a straightforward statement of the performance requirements without any substantive concept design work could be placed in this category. In this case the basis of the contractors appointment would contain little or no pre-design work. The main features of this approach are:

- It allows an early indication of cost certainty
- Installation quality standards will probably be the minimum practical
- It will be difficult to compare tenders as there is no common design basis
- The contractor will have all the design responsibility
- It is really only practical if the rest of the project is procured in the same manner; if the architectural and structural designers are on full duties the building services design and build contractor has two potentially contradictory roles, i.e. designer and installer

Most design and build projects require more information than a simple statement of the performance requirements. Frequently the project design development is taken to the end of the scheme design before the building services work is tendered. The main features of this approach are:

- It allows the designer to integrate the bulk of the design requirements into the project before the contractor is appointed
- Tenders can be compared on a common design basis providing a 'yardstick' for cost comparison
- It provides a level of design certainty, i.e. there is more control over the design outcome
- The installation quality can be improved through knowing the design requirements and issuing the appropriate installation specifications
- Design responsibility will be shared between the designer and the contractor; the scheme design will be the designers responsibility and subsequent design work the responsibility of the contractor; in situations where the scheme design package includes indicative plant and equipment selections the designer may ask the contractor to review the selection in conjunction with the detail design development

Irrespective of the type of approach taken with design and build contracts (like installation contracts) the tender return only provides a fixed cost for the work currently described in the contract documents, installed in a set sequence. It is very likely that this vision of the work requirements will not actually occur between tender and contract completion and consequently costs will increase. The solution is to allocate a project specific risk budget on top of the tender cost. The size of the risk budget needs to be discussed with the project team and specifically the quantity surveyor.

7.9.2 Allocating design responsibility

Allocating design responsibility and ensuring it is accepted and understood across the project team is often difficult. The issue is common to all project work as the design work is invariably split between the designer and the contractor. The level of design ownership varies with the procurement route. Simple performance design deliverables place most of the design responsibility on the contractor while detail design deliverables place most of the design responsibility on the designer. Alternatives between these two extremes vary the design responsibility of both parties. The key issue is to clearly define who is responsible for each of the design elements and ensure this is compatible with the contractor's form of contract and the designer's conditions of engagement.

A method for improving clarity in communicating design responsibility is discussed in Ref. [6]. In summary the design activities and deliverables are explicitly defined, e.g. schematic drawings, co-ordination drawings, operation and maintenance manuals, etc.

Each of the activities and deliverables are presented in a design responsibility matrix and allocated to the designer, the installer or other (design and build contractor, specialist subcontractor, etc). The publication presents a set of design deliverable definitions and a design responsibility matrix pro forma for general use. These documents and an interface matrix (i.e. a list of all the building services interfaces and who is responsible for the final connections) must form part of the design deliverable package.

7.10 Chapter review

This chapter explains the process and deliverables involved in scheme design through the following topics:

1   The design process
2   Developing the design information
3   Design work
4   Value engineering
5   CDM requirements
6   Risk assessments
7   Scheme design deliverables:
    • Technical performance specifications
    • Scheme design drawings
8   Design responsibility

Completion of scheme design finalises the design development work and provides a platform for design detailing.

References

[1] *Value Engineering of Building Services*, Building Services Research and Information Association Publications, Application Guide 15/96.
[2] *Designing for Health and Safety in Construction*, HSE Books.
[3] *National Engineering Specification*, Barbour Index, Ltd. Publications.
[4] *Design Information Flow*, Building Services Research and Information Association Publications, Technical Note 17/92.

[5] *Guide to the Preparation of Specifications*, BSI Publications, 1991, BS 7373.
[6] *Allocation of Design Responsibilities for Building Engineering Services,* Building Services
Research and Information Association Publications, Technical Note TN 21/97.

Further reading

Guide to the Preparation of Specifications, BSI Publications, 1991, BS 7373.
A Guide to Managing Health and Safety in Construction, HSE Books.

Chapter 8

Detail design stage

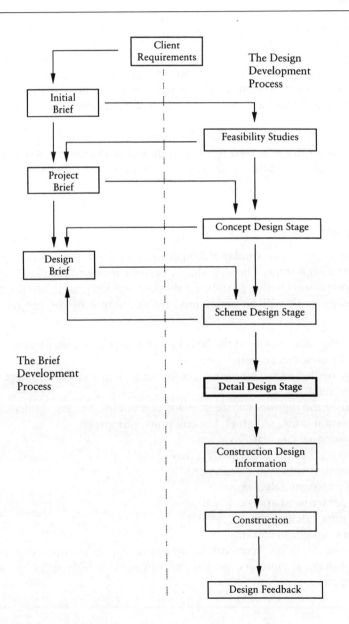

8.1 Chapter contents

The topics covered in this chapter include

- Design activities
- Design margins
- Detail design good practice
- Design deliverables

8.2 General

The detail design process is essentially an enhancement of the design work carried out in the scheme design stage. It should not involve any evaluation and development of design information or concepts, if the previous design stages are correctly managed and 'closed out'. However, late design changes do occur and in these situations the affected section of the design needs to be reworked through the concept or scheme design activities (or both) and then revised, re-issued and signed off.

Detail design work only applies to HVAC and utility services systems. Fire protection work is generally procured through technical performance design deliverables and noise control work through the services of an acoustic consultant (see Section 7.8.9).

8.3 Design activities

Detail design activities are similar although more in-depth than the design activities in the scheme design stage. The size, shape, performance and requirements of all the individual components will need to be established and integrated into the design. The range of design work will be determined by the nature of the project and could include:

- Confirmation and review of the heat loss and heat gain calculations
- Pipework sizing calculations
- Selection of the inline pipework components, e.g. valves, strainers, air vents, commissioning sets, etc.
- Selection of the pipework systems plant and equipment, e.g. boilers, calorifiers, pressurisation units, radiators, fan coil units, pumps, etc.
- Air conditioning calculations
- Room pressurisation and leakage calculations
- Estimated particle count calculations
- Fugitive emissions calculations
- Smoke extract ventilation calculations
- LEV capture calculations
- Ductwork sizing calculations
- Selection of the inline ductwork components, e.g. balancing dampers, constant volume boxes, attenuators, grilles and diffusers, fire dampers, pressure relief dampers, etc.

- Selection of the ductwork systems plant and equipment, e.g. air handling units, filters, coils, fans, humidifiers, run around coils, dust extract unit, etc.
- Energy consumption calculations
- Electrical load calculations
- Pipework and ductwork insulation selection
- Co-ordinated integration of the proposed design into the building(s)
- Preparation of the detail design schematics, standard detail sheets, plan and section layouts and plant room drawings
- Production of plant and equipment schedules, e.g. grilles and diffusers, commissioning sets, control valves, etc.

The correct format for calculations is important as they will need to be quickly and easily understood by a range of people with different requirements and varying levels of knowledge (see Section 7.4.1).

8.4 Design margins

Design margins are 'custom and practice' within engineering and are invariably established from past design experience. Unfortunately there is little authoritative reference on the actual margin to be applied in any specific application. An article in the CIBSE magazine 'Design Margins in Building Services' August 1997 based on an industry survey, summarises the considerable range of design margin values used in similar applications. More recently liberal use of margins has been blamed for the over engineering of building services systems.

Although there are good reasons for applying design margins (future use requirements, installation tolerances, theoretical fabric design U values, etc.) any objective figure can be debated. Based on past project experience and where appropriate from reference texts a list of applications and design margins is provided in Table 8.1. *These may provide a reasonable starting point for design calculations although the final decision on the application of a design margin and its value remains with the designer.*

8.5 Detail design good practice feedback

There is a considerable amount of literature providing technical detail design guidance and the aim of this section is not to repeat what is already available. (General information on design guidance is provided in Section 5.4.2, and specific guidance in the further reading section of this chapter). This section feeds back some past project experience and provides an indication of acceptable good practice.

8.5.1 General detail design good practice

The following general detail design good practice issues cover a wide range of different subject areas. To assist quick referencing they have been separated into the following broad categories: general, ventilation, pipework, plant and equipment.

The decision to apply the information described below is the responsibility of the designer.

Table 8.1 Design margins

Application	Design margin or criteria
Heat loss calculations	10% on the heat loss calculations
Pipe sizing calculations	10% flow rate addition to the 'marked up' heat loss calculations to allow for heating mains losses
	10% flow rate addition to the heat gain calculations for chilled water mains losses
	Alternatively calculate likely mains losses
Radiator selection	No addition to 'marked up' heat loss calculations
Heating and cooling coils	No addition to calculated fluid flow rates
Boiler selection	Use pipe sizing calculations for initial boiler load
	Increase proposed margins for preheat requirements
Pump selection	115% of the design flow rate against the design system pressure
Heat gain calculations	No mark up on calculations
Fan selection	110% of the design flow rate against 115% of the design system pressure
Air cooled package chiller selection	Use pipe sizing calculations for initial chiller load
	Ensure external ambient design temperatures are consistent with current actual temperatures and process needs in addition to CIBSE air conditioning recommendations
	Ensure pump heat gains into the closed loop system have been allowed

8.5.1.1 General

- Prior to detail design of air and water systems review the existing scheme design against the recommendations in BSRIA Application Guide 3/89.2 The Commissioning of Air Systems in Buildings and BSRIA Application Guide 2/89.2 The Commissioning of Water Systems in Buildings.
- Subject to the clients' agreement design clean areas for upgrade to one classification higher than the current intended use. Most clean processes are likely to be upgraded as GMP and hygiene standards continue to increase.
- Prior to any detail design review the CIBSE Application Manual 'Automatic Controls and their Implications for Systems Design' 1985 [1] to establish how the system will be controlled before it is designed. For instance, a multiple reciprocating chiller or boiler plant installation should be controlled with a return detector. (Constant flow control under part load conditions could lead to subcooling on the chiller installation and over-temperature on the boiler plant.)
- Final selection of grilles, diffusers, commissioning sets, steam system ancillary items, attenuators, boilers, pumps, chillers, fans, etc. should be made by the proposed manufacturer with a model reference.
- The range of new non-ozone depleting refrigerants have different characteristics. Currently the most suitable applications for standard HVAC work are R407C for small split dx units and drop-in replacement for R22 (subject to the equipment being suitable), R134A for new chillers designed for this refrigerant and R404A for applications less than 5°C, e.g. cold rooms.
- Ensure the building fabric design is suitable for fixing and supporting the build-

ing services systems. Steel decking systems (with cast concrete slabs) can be provided with universal fixing systems (e.g. a twisted wedge with a drop rod fixing), brick and cast concrete are also acceptable. Pre-stressed concrete structures should be avoided (particularly pre-stressed concrete T beams with hollow infill pots).

- Roof structures and decking systems should allow adequate easy access for maintenance requirements and likely future use equipment loadings (e.g. split dx condenser units, chillers, air handling equipment, etc.).
- Where there is a suitable heat sink, e.g. 'warm' process water storage, consider direct heat reclaim from air cooled chillers. Manufacturers can supply air-cooled chiller packages with supplementary coils for heat reclaim.
- Provision for maintenance down time and future load requirements translates into specific plant and equipment design requirements. This manifests itself in different options for different systems and different applications. Common plant and equipment provisions are listed below although they are only for general guidance and will need to be adjusted to suit the specific project requirements:
  - Three chillers with 50% design capacity each, for process critical chilled water applications, e.g. confectionery production area
  - Two chillers with 60% design capacity each, for non-process chilled water applications, e.g. office air conditioning
  - Two boilers with 60% design capacity each, for non-critical process heating applications
  - All pumped systems (except domestic hot water systems) should have duty and standby pumps (single body, twin head or two single pump sets)
  - Provide duty and standby fans (with separate drive shafts) on all process critical ventilation systems
  - Provide double block and bleed valve arrangements for isolation on all steam systems
- Ensue that a comprehensive description of the commissioning requirements is detailed in the specification. This should include the appointment of an independent commissioning company and a clear responsibility on that company to ensure that all pre-commissioning checks are made prior to commissioning the project works (see Section 10.6).
- Ensure that the operation and maintenance requirements include the appointment of an independent documentation specialist to co-ordinate and compile the documents. Also consider the document form, hard copies of documents will need to be supplemented with back up CDs.

8.5.1.2 Ventilation

- Where product specific capture velocities are not available LEV design work should be based on HS(G) 37 Introduction to Local Exhaust Ventilation and the Industrial Ventilation Manual of Recommended Practice.
- Hazardous areas require a minimum ventilation rate to prevent any concentration of flammable gases and an increase in hazardous rating beyond the design basis. Although most hazardous rating design documentation recognise this requirement

none provide a definitive figure for this type of application. In the absence of other design information consider a minimum five air changes per hour (taken from HS(G) 51 The Storage of Flammable Liquids in Containers, 1990). *However the only method of ensuring a suitable environment is to carry out fugitive emissions calculations and design the fresh air ventilation rate to maintain the environment significantly below the lower explosion limit (LEL).*

- Design of LEV and VOC extract systems should be based on all the extract points operating at the same time. No diversity should be applied to the system design unless the operational methodology is known and the limitations understood by the client.

- When design requirements necessitate a full fresh air or single pass ventilation system consider installing an air handling unit with a built-in run around coil section rather than an independent run around system.

- Where HVAC systems are designed to create pressure regimes, consider providing a supply and extract point in each space and a pressure relief damper between adjacent interconnected spaces. The supply and extract provides a degree of commissioning flexibility when establishing the actual leakage from the space (if the system design includes constant volume boxes ensure that they incorporate easy 'on-site' adjustment). The pressure relief damper provides a visual demonstration that a pressure regime exists, it also compensates for 'wear and tear' reduction in the door seal. Where a design face velocity is required across a door the pressure relief damper can pass the design air volume (less the door leakage) when the door is closed.

- In complex HVAC designs, e.g. a pressurised dispensary suite, ensure that the space conditioning is split up into areas served by a single air handling system, so that the air supplied in one space that leaks into another space is extracted by the same air handling system. This prevents a mismatch between air handling systems, e.g. one system supplies 10 m^3/s and extracts 4 m^3/s while the adjacent unit supplies 10 m^3/s and extracts 16 m^3/s (dumping 6 m^3/s to atmosphere).

- Large commissioning variations in air volumes to pressurised spaces can result in significant coil demand changes. Where this is likely to occur change the construction fabric details to control and define the leakage and add a suitable design margin to the coil selection.

- Chilled water (5°C flow, 11°C return) contra flow cooling coils will have a mean coil surface temperature of approximately 9°C (the actual figure should always be confirmed with the supplier). Allowing a design margin of 1°C, supply air conditions of less than 10°C are not practical with this cooling medium. (This is particularly apparent in low humidity applications.) Applications requiring low temperatures should consider refrigerant or inhibited water solutions, i.e. brine or propylene glycol mix.

- Air conditioning design for wet process areas should include a latent load for wet washdown of the process and evaporation into the atmosphere. The differential vapour pressure between the design environment and the washdown environment can be determined from the CIBSE Guide C1, Properties of Humid Air, Water and Steam.

- Backward curved fans impellers are preferable to forward curve fan impellers as they are self-limiting (power absorbed falls after a peak) and cannot be over-loaded. This is particularly useful during commissioning if the system pressure demand is greater than the anticipated design and the fan rpm needs to be increased.

- UK GMP guidance (and the available new ISPE baseline guides) only specify 20 air changes an hour for sterile suite clean areas. All other clean area applications do not require this level of ventilation although misunderstanding and concern over the view of regulatory bodies often prompts this solution. The main practical drivers are the particle count and the clean-up rate in the space.
- Containment booths can be installed as single pass or recirculation systems. Single pass booths are used on liquid handling or any similar applications that emit hazardous fumes, e.g. solvents. Solids handling applications can use recirculation systems as the dust product is held in the filters. Recirculation booth installations are approximately 60 per cent cheaper than single pass booths as no supply or extract ductwork is required and their design OEL is the same as the single pass booth. However, recirculation booth installations do not dissipate the significant fan power gain generated by creating hundreds of air changes an hour within the enclosed booth safe operating area. This type of installation should be provided with a supplementary cooling coil to mitigate the heat gain.
- Ventilation or air conditioning systems serving complex pharmaceutical applications are often provided with invertor drives and constant volume boxes prior to terminal HEPA filters. Invertor drives provide flexibility during commissioning and ensure that adequate pressure is available to operate the constant volume boxes. The constant volume boxes mimic the clean/dirty terminal HEPA filter requirements and provide a measure of air volume control (in conjunction with the invertor drive) when doors to the suite, etc. are left open. The invertor drives must be controlled to maintain a static pressure in the duct, allowing the fan to respond to changes in the system characteristic. Velocity control of the invertor drives in this application is *not* acceptable as the fan system (controlled by velocity) can fight the constant volume boxes (controlled by pressure).
- Where ventilated spaces require the introduction of a large number of air changes normal air diffusers may overthrow and create draughts. Alternative solutions are swirl diffusers, plenum ceilings (with 'drop-in' perforated plates) and air socks.
- Where designs involve adjacent change and shower/toilet areas, consider supplying air into the change room and extracting the same air through the shower/toilet reducing the total air volume.
- Where ventilation systems are being used to control space pressure by using a defined supply or extract volume always consider the effect of external wind pressure. Changes in wind pressure can effect the air handling system volume and space pressure.

8.5.1.3 Pipework

- In process applications that distribute steam and require hot water for washing down consider the use of direct steam heated mixing valves that are served with cold potable water and produce instantaneous hot water.

- LTHW heating systems are traditionally designed for 82°C flow and 70°C return. When the LTHW system incorporates a pressurisation unit (i.e. adequate static head) consider 90°C flow and 70°C return – increasing the emitter average temperature and reducing the flow rate.

- On condensing boiler LTHW systems that only serve air heater batteries consider 50°C flow and 38°C return (or a similar return temperature below the condensing temperature of the combustion gases) as this will ensure permanent condensing in the boiler and maximum efficiency. In these situations the nominal increase in coil cost for operating at 50°C/38°C will be quickly paid back in energy savings.

- Where a LTHW condensing boiler serves a range of radiators, use direct compensation control to increase the operational boiler efficiency. If the project involves air heater batteries consider adding them to the compensated circuit and revise the coil selection to suit the new operating criteria. (Check with the coil manufacturer that the coil LTHW supply conditions can satisfy the coil output over the full range of operating conditions.)

- Steam to LTHW calorifiers or heating coil installations should include an anti-vacuum device on the condensate line prior to the trap installation and an air outlet device (e.g. a thermodynamic trap) on the steam line at the battery. In part load conditions or when the supply valve closes the remaining steam in the battery collapses pulling a vacuum and closing the condensate trap. The anti-vacuum valve allows air to be drawn into the battery and unseals the trap, freeing the stored condensate. When the supply valve opens (on demand) the air outlet device allows the trapped air to be removed and replaced with steam.

- Condensate lines should be designed to fall away under gravity from the heating appliance. In situations where this is not possible (i.e. an existing installation) a pumped trap should be installed, allowing the condensate to be pumped back into the condensate main at a high level. If a pumped trap is not provided the static head of condensate in the pipe rising back up into the condensate main will inhibit the operation of the steam supply valve. For instance in a cold start or part load condition the steam valve will first have to open wide enough to develop enough pressure to shift the residual condensate (with a great deal of water hammer) before it can address the heat load demand.

- The containment of dip pipe lance emissions when dispensing liquid chemicals from drums is significantly reduced with a drumvent dip pipe sleeved enclosure (the dip pipe is enclosed within a vented sheath).

- Pumped chilled water pipework systems are subject to a system heat gain from the work the circulating pump is adding to the closed circuit. This effect is commonly recognised in ventilation systems as fan power gain and applies equally to pumped pipework systems. The heat gain to the system (kW) is equivalent to the system pressure drop (kPa) multiplied by volume flow rate (m^3/s). For example, a chilled water flow rate of 0.15 m^3/s (150 kg/s) with a system pressure drop of 110 kPa will add a heat load (and consequently an additional cooling duty) to the system of 16.5 kW.

8.5.1.4 Plant and equipment

- Reciprocating air cooled liquid chillers that have a limited number of 'on/off' switchings per hour will require chilled water buffer vessels, if the chilled water pipework system does not have adequate thermal storage capacity.
- The choice of traditional reciprocating or screw chiller selection should be considered against the benefits of gas fired (or site steam) absorption chillers. Primary energy requirements and CO_2 emissions are lower than traditional chillers. Integration of absorption chillers with CHP plants provides significant energy saving benefits
- Air handling units benefit from internal lights, viewing ports, space for instruments (e.g. after the frost coil), access sections and adequate space between cooling and heating coils to prevent moisture carry over (close coupling also reduces the coils performance)
- Where practical specify the 'on' and 'off' AHU conditions rather than the coil conditions. Manufacturers are better placed to select the most cost efficient unit if they are given the overall design requirements rather than the component requirements
- When selecting pumps for open systems, e.g. pumps returning water from cooling tower reservoirs or water storage tanks, always ensure the pump net positive suction head (NPSH) is adequate

8.5.2 Technical detail design good practice

The following list of technical detail design good practice issues covers typical building services subject areas. *The decision to apply the information described below is the responsibility of the designer.*

- Minimum external, fresh air supplied per person in occupied areas should be 12 l/s. Other design fresh air requirements should be taken from CIBSE Guide Table A1.5
- Velocities in ductwork under normal conditions should be designed between 2.5 (final connections) and 7.5 (main duct runs) m/s
- Ductwork pressure drops should be between 0.6 and 1.2 Pa/m
- Face velocities on cooling coils shall not exceed 3.0 m/s unless eliminator plates are installed after the cooling coil
- Face velocities on heating coils shall not exceed 3.5 m/s
- Pipework velocities should be a minimum 0.5 m/s. Maximum pipework velocities should be 1.5 m/s up to and including 150 mm diameter pipe and 2.5 m/s for larger pipe sizes
- Chilled water and LTHW pipework pressure drops should not exceed 300 Pa/m.
- Steam pipework velocities should not exceed 15 m/s
- Steam condensate pipework should be sized for twice the maximum design flow rate to allow for cold start-ups
- Constant volume valves (e.g. autoflow or similar) and suitable variable speed pumps should be considered for use on closed loop pipework systems
- Building services control systems controlled by ddc outstations should be provided with at least 10 per cent spare digital inputs and 10 per cent analogue outputs

- Commissioning station orifice plate selections should be within a design pressure drop range of 1–8 kPa (this is the economic selection of head loss against capital cost of the device and is also the measurement range of traditional water based manometers)
- clear access should be provided throughout plant areas; the *minimum* clear opening from finished floor to underside of any obstruction should be 2.2 m

8.6 Value engineering

On large projects, where insufficient detail is available at scheme design to comprehensively value engineer the design solution, a further review may be required during the detail design stage (see Section 7.5).

8.7 CDM requirements

The general safety hazards have been identified in the concept design and recorded with indicative solutions in the design brief. Scheme design takes the initial work and refines the general solutions into specific answers. During the design detailing process further specific hazards are likely to be identified and the review process established in the scheme design stage will need to be repeated (see Section 7.6).

After completion of the detail design CDM review any 'significant' hazards inherent in the final design solution must be recorded and passed to the planning supervisor. The planning supervisors duties involve confirming all the project hazards to the principal contractor through the health and safety file. Frequently it is useful to include the hazard list in the technical specification appendix.

8.8 Improving construction productivity

The Egan Report 'Rethinking Construction' set seven key targets for the UK construction industry to achieve world class construction best practice. The targets are:

- Reducing capital cost by 10 per cent
- Cutting construction time, from client approval to practical completion, by 10 per cent
- Increasing by 20 per cent the number of projects completed on time and within budget
- Reducing by 20 per cent defects on handover
- Cutting back on reportable accidents by 20 per cent
- Increasing the value-added productivity (per head) by 10 per cent
- Increasing turnover and profits of construction firms by 10 per cent

'How' the targets will be achieved is still being evaluated through various industry focus groups. The immediate impact on designers' work is a shift towards standardisation and enhanced constructability. Standardisation has two aspects, standard design solutions and physical standardisation of plant and equipment. Although this is a complex area involving all the parties in the construction process, listed below are a number of issues that designers may want to consider in an attempt to improve site productivity. (In the feasibility and concept design stage these issues form part of the constructability considerations.)

8.8.1 Design considerations

- Use of standard design solutions e.g. BSRIA Library of System Control Strategies, BSRIA Standard Specification for BEMS
- Overall design 'constructability', e.g. ease of access, use of push fit connectors
- Prefabrication, e.g. fan coil and AHU sets, plantroom modules, toilet modules, pipework risers
- Cable management systems
- Universal fixing systems
- Standardisation of plant and equipment, e.g. design based on a standard fan coil unit
- Material selection, e.g. steel, copper, uPVC, ABS, sheet metal ducts, fabric ducts

8.8.2 Improved design solutions

- Variable speed pumps and constant volume valves
- Polyethylene underground pipework systems
- Direct hot water from steam, cold water mixing valves
- Inflatable ductwork (permeable or impermeable with outlets)
- Prefabricated fan coil unit sets
- Drumvent dip pipes for liquid handling problems
- LTHW design temperature, 70–90°C
- Recirculation containment booths (with cooling coils)
- Full fresh air AHUs with built-in run around coils

8.8.3 Constructability

- Quick connection joints, 15–50 mm (rather than screwed)
- Victolic joints, >50 mm (rather than welded)
- Flexible pipe connections with quick connectors
- Increased use of flexible ductwork
- Skid mounted equipment, e.g. boiler and chiller sets with pressurisation units
- Integrated package control equipment, e.g. TREND local outstations on AHUs
- Wire hangers (up to 315 diameter ductwork)
- Universal fixing systems
- Pre-insulated ductwork and pipework

Further useful suggestions on standardisation are usually available from equipment manufacturers and contractors (who will have an increasing influence on the design process).

8.9 Detail design deliverables

Detail design deliverables only apply to HVAC and utility services work. Generally fire protection work is detailed in a technical performance specification and noise control work through the acoustic consultants supplements to the project design deliverables.

The issues to consider before starting the design deliverables and the structure of specifications are the same for detail design deliverables as scheme design deliverables (see Sections 7.8.2 and 7.8.3).

8.9.1 HVAC and utility services technical detail design specifications

Detail design specifications describe the specific technical requirements of the project work. They do not contain any contract preliminaries or any installation requirements.

The main section headings within any detail design specification are always the same as they prompt the type of detail that applies to all projects. The primary section headings are:

- Scope
- Design description
- Project specific requirements
- Automatic controls installation
- Acceptance criteria
- Equipment schedules

The remainder of this section describes the generic content of each of the main section headings.

8.9.1.1 Scope

The aim of this section is to provide a brief introduction to the project and describe the proposed works. The section comprises of the following subsections.

(a) *Introduction*: a short paragraph describing the type of work, the client and the site address, e.g. 'This detail design specification provides the basis for the construction design information and installation of the...(HVAC or utility services) systems and equipment for...(client name)...located at...(and site address).'

(b) *Scope of services*: a list of the services that the contractor provides as part of the contract works, e.g.

- Provision of construction design information deliverables
- Provision of sample materials and off-site material testing
- Provision of the scope of work requirements
- Progress meetings in the design organisation offices, etc.

(c) *Scope of work*: a list of the main HVAC and utility services systems and all associated services enabling work required within the contract works, e.g.

- A high pressure steam and condensate pipework installation to the manufacturing plant
- A LTHW and chilled pipework system to the office building fan coil system

- A primary, multiple air cooled chiller plant installation
- A ddc control system to the HVAC and utility services systems
- Commissioning of the all HVAC systems, etc.

8.9.1.2 Design description

The aim of this section is to ensure that the contractor fully understands the HVAC or utility services system requirements. This section describes each of the scope of work items listed in the previous section.

8.9.1.3 Project specific requirements

The project specific requirements section involves describing any unique project requirements that impact on the detail design and explaining the extent of the construction design information deliverables.

Unique project requirements that impact on the contract works rather than just the HVAC and utility services typically include issues such as GMP, validation, food hygiene requirements, containment, etc.

This section should include a list of all the construction design information deliverables (including the number of copies) the contractor needs to produce to ensure the design intent requirements have been satisfied, e.g.

- Ductwork installation drawings (1:50 scale)
- Ventilation and pipework systems schematics (NTS)
- Plantroom layout installation drawings (1:20 scale)
- Pipework installation drawings (1:50 scale)
- Construction detail drawings for roof and wall penetrations (NTS)
- Manufacturers plant and equipment drawings, including control panel and controls wiring drawings, etc.

8.9.1.4 Automatic controls installation

This section of the specification details the automatic controls installation associated with the contract works. The detailed requirements in conjunction with the schematics and installation standards (these should be included in the installation specification) will provide the specialist controls subcontractor with a comprehensive understanding of the controls' requirements.

Guidance on specifications for BMS and controls installations is provided in Refs. [1–5].

The individual project requirements determine the precise content of this section, however, as a minimum it should contain the following subsections.

(a) *Normal operation description*: a 'normal operation' description of the individual systems including start/stop times, temperature and humidity set points and control bands, compensation control, optimisation, hazardous area requirements, dust zoning requirements, pressure regimes, etc. with cross-referencing to the schematic drawings and any relevant tabulated data, i.e. the design criteria presented in the design brief document.

(b) *Emergency operation description*: a description of the systems operation in emergency conditions, for example fire alarm conditions, fail safe shutdown requirements, containment failure alarms and procedures, emergency spill ventilation requirements, etc.

(c) *Controls interfaces*: details of the interfaces between the new controls installation and package equipment, other subcontract packages, existing systems, etc.

This section also includes listing the free issue items that the controls contractor will supply to the contractor.

(d) *Control panel description*: this section describes the number of control panels and the generic panel fascia mounted equipment. Fascia mounted equipment typically consists of:

- Hand-off auto rocker switches for time control plant
- Hand-off rocker switches for package plant equipment and plant items that are not time controlled
- Run (green), trip (red) and panel live (white) lamps
- BMS key pad consoles
- Door interlocked isolator
- Package plant common fault alarm lamps
- Fire alarm reset push buttons and key switch override
- Hours run indicators, ammeters, voltmeters, etc.

(e) *Testing and commissioning requirements*: a list of the testing and commissioning services the controls contractor is expected to provide. This would typically include:

- Controls installation wiring acceptance checks and where specified final connections to the control panel, plant and equipment
- Integrity testing of the power and controls installation
- Site attendance during the commissioning period to support the contractor and the commissioning specialist
- Setting to work and commissioning of the controls installation
- Supply of the commissioning documentation for inclusion in the O&M documentation
- Proving of the controls installation to the site engineer
- Performance testing of the overall controls system and where applicable the involvement in the acceptance testing requirements
- Post-installation maintenance and remote installation monitoring during the defects period

(f) *General*: this section includes any item of work that is specific to the project and not readily incorporated into any of the other sections. This may include:

- Cross-reference to the provision of construction design information deliverables listed in the project specific requirements section of the specification
- Attendance at meetings on site or in the design organisation's offices
- Software validation requirements
- Training of the clients staff
- Visits to existing controls installations

8.9.1.5 Acceptance criteria

The aim of this section is to explicitly state the acceptance criteria for the contract works. The section will repeat selected key acceptance criteria agreed with the client in the design brief document and where necessary explicitly define other acceptance criteria, e.g. acceptable commissioning flow rate tolerances, design noise rating levels in the occupied space, etc. Any additional acceptance criteria which needs to be specified in the format described in the design brief section on key acceptance criteria (see Section 6.11).

8.9.1.6 Equipment schedules

This section contains all the schedules that describe the contract works plant and equipment.

The schedules should provide enough information to adequately describe the plant and equipment. Invariably the contractor will use the schedules to obtain tender quotes and, if successful, to order the plant and equipment. Consequently the schedules should precisely describe the plant and equipment and include the preferred manufacturers' name, address and telephone number. They should not contain redundant information, information repeated elsewhere in the specification (i.e. they should be complete and stand alone) or any reference to supplier quotes.

Contractors and suppliers have often been 'caught out' when tendering for work as the contractor contacts the supplier during the tender stage and the supplier quotes the cost for the plant or equipment based on the initial discussions held previously with the designer, without viewing the tender schedule. Meanwhile the design requirements have changed and the supplier has underestimated the cost or the contractor provides the wrong plant or equipment. Consequently the schedule should include a footnote to prompt the contractor to ensure that all the necessary information has been sent to the supplier.

8.9.2 Installation specification for HVAC and utility services work

The installation specification details the installation quality requirements for the systems described in the detail design specification. The broad nature of building services means that these types of documents are very large, as they endeavour to cover every installation aspect of every system. The master installation specification is reviewed by the designer and editted to ensure that:

- The systems proposed in the contract works are comprehensively and correctly detailed
- The installation requirements proposed are in accordance with the specific project requirements
- The sections of the document that are not relevant are removed to avoid confusion

The content of master installation specifications are built up over a considerable number of years with the experience of each project fed back into the document.

Generally they are the copyright property of the design organisation and are not repeated in this text; however, the contents typically include:

- Installation standards
- Pipework services
- Ductwork services
- Thermal insulation
- Testing and commissioning
- Operating and maintenance documentation

Note: this section is repeated in Section 7.8.5.

8.9.3 Detail design drawings

The detail design drawings are a method of communicating the physical detail behind the design intent, described in the detail design specification.

General guidance points for preparing detail design drawings include:

- Provide only as much detail as is necessary to communicate the design intent.
- Each detail design schematic should be a summary of the detail information for that section of the work. In addition to normal schematic plant and equipment requirements the drawings should include all automatic controls instrumentation, design flow rates and velocities, design operating requirements, i.e. temperatures, humidity, particle counts, etc., all control valves and dampers, all isolating and commissioning valves, measuring stations, sample points, access requirements, etc. The final drawings should contain enough design information for a competent designer to understand the design intent without reference to any other information.
- Before completing any schematic drawings ensure that a mass and temperature balance has been carried out.
- Where information is indicative ensure that the drawing contains an obvious note stating the 'the information on this drawing is indicative only' or similar.
- Pipework and ductwork layouts should be 1:50 scale. Where appropriate they should be dimensioned from the design grid with reference to a practical setting out point.
- Plantroom drawings should be 1:20 scale with layout and section information.
- Standard design details, e.g. commissioning sets, coil details, sleeved penetrations through walls, fan installations, etc. should be available from successful previous projects.
- Provide dimensioned information where necessary.
- If an area contains services detailed by other disciplines ensure all their appropriate drawings are cross-referenced on the building services layouts.

8.10 Chapter review

This chapter explains the detail design process through the following topics:

1 Design activities
2 Design margins
3 Detail design good practice feedback
4 Improving construction productivity
5 Detail design deliverables:
   a Detail design specifications
   a Detail design drawings

The end of the detail design stage is followed by the tender period, contractor interviews and appointment of a contractor. The design detailing work is complete. The designer's activities now shift from design work to policing and design verification.

References

[1] *Automatic Controls and their Implications for Systems Design*, CIBSE Application Manual, 1985.
[2] *Library of System Control Strategies,* Building Services Research and Information Association Publications, Application Guide AG 7/98.
[3] *Guide to BEMS Centre Standard Specification*, Volume 1, Building Services Research and Information Association Publications, Application Handbook AH 1/90.
[4] *Standard Specification for BEMS*, Volume 2, Building Services Research and Information Association Publications, Application Handbook AH 1/90.
[5] *Specifying Building Management Systems,* Building Services Research and Information Association Publications, Technical Note TN 6/98.

Further reading

Guide to the Preparation of Specifications, BSI Publications, 1991, BS 7373.
Allocation of Design Responsibilities for Building Engineering Services, Building Services Research and Information Association Publications, Technical Note TN 21/97.
Designing for Health and Safety in Construction, HSE Books.
The Commissioning of Water Systems in Buildings, Building Services Research and Information Association Publications, 1998, Application Guide AG 2/89.2.
Pre Commissioning Cleaning of Water Systems, Building Services Research and Information Association Publications, Application Guide AG 8/91.
Water Treatment in Building Services Systems, Building Services Research and Information Association Publications, Application Guide AG 2/93.
The Commissioning of Air Systems in Buildings, Building Services Research and Information Association Publications, 1998, Application Guide 3/89.2.
Guide to the Selection and Installation of Compressed Air Services, British Compressed Air Society, 1992.
Services Supplying Water for Domestic Use within Buildings and their Curtilages, BSI Publications, 1987, BS 6700.
Heating Control in Large Spaces, Building Services Research and Information Association Publications, Technical Note 23/97.
Oversized Heating Plant, Building Services Research and Information Association Publications, Guidance Note 12/97.

Condensing Boilers, Building Services Research and Information Association Publications, Technical Advice 1/90.

Designing for Maintainability, Building Services Research and Information Association Publications, Application Guide 11/92.

Prefabrication and Preassembley – Applying the Techniques to Building Engineering Services, Building Services Research and Information Association Publications, Advanced Construction Techniques 1/99.

Chapter 9

Construction design information

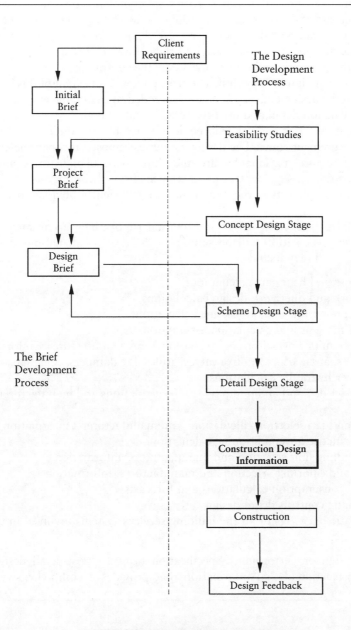

9.1 Chapter contents

The topics covered in this chapter include:

- Construction design information
- Content of construction design information
- Managing construction design information

9.2 Construction design information

The designers 'recognised' design activities are complete either at the end of scheme design or the end of detail design, depending on the procurement route. From the time the contractor is appointed the designers' role changes from carrying out design activities to policing and design verification of the contractor's works.

Prior to the start of installation work on site the contractor needs to provide a range of construction design information. The type, content and extent of this information is described in the design deliverables and is intended to ensure the engineering scope of work, construction detail and quality are fully understood.

The designers' work and role will be determined by the conditions of engagement and the procurement route. The designers' duties imposed through the conditions of engagement require 'reasonable care and skill' to be demonstrated and for work designed by others, e.g. subconsultants, specialist subcontractors, etc. 'to ensure that there are no defects in the design that are sufficiently obvious that they should have been spotted'.

The initial construction design information supplied by a contractor through a performance specification will be similar to the design calculations undertaken in detail design and may include:

- Heat loss and heat gain calculations
- Pipework and ductwork design calculations
- Room pressurisation and leakage calculations
- Design margins and calculation assumptions
- In line equipment selection calculations and manufacturer's information, e.g. radiators, attenuators, control valves, grilles, fire dampers, coils, safety valves, etc.
- Sprinkler hydraulic calculations
- Fire water demand and pump selection calculations and manufacturer's information
- Pump and fan selection calculations and manufacturer's information
- Psychrometric air conditioning calculations
- Air handling unit selection and manufacturers' information
- Boiler and calorifier selection and manufactures' information
- Energy consumption calculations and forecasts
- Automatic controls detail design
- Schematic drawings of all the building services systems involved in the contract work, etc.

Detail design or performance specification projects 'post-detail design' need to supply construction design information that provides a comprehensive picture of the works before installation. Typically construction drawing information would include:

- Ductwork layout construction design drawings
- Ventilation and pipework systems schematics
- Plantroom layout construction design drawings
- Pipework layout construction design drawings
- Construction detail drawings, e.g. roof and wall penetrations, fire damper installation details, pipe and duct support details, plinth details, etc.
- Manufacturer's plant and equipment drawings for all the plant and equipment supplied in the contract works
- Control panel drawings, wiring drawings and schedules of instrumentation
- Co-ordination drawings detailing the inter relationship between the building services systems, the building fabric and the other discipline services in the vicinity, etc.

9.3 The content of construction design information

In principle, the construction design information needs to contain enough detail to allow a competent person, not involved in the contact works, to interpret the design and installation requirements. This level of detail is obviously different than the minimum amount of information the contractor needs to produce to install the works. Although the precise construction design information requirements may have been explained in the specification it is usually beneficial to highlight the requirements again to the contractor before or when the first set of information is produced.

The designer's task is to review and check the information for compliance with the design intent. Although all project work is different a summary of the type of 'checking' work is detailed below.

9.3.1 Construction design information – general

- Design calculations should be audited as if they were produced by the design organisation. The design margins and design good practice detailed in the performance specification should have been applied across all the design work. Calculations should be checked for mathematical correctness and perhaps more importantly for design assumptions.
- All construction design information should be suitably documented. The information needs to be indexed, drawings need a title, scale and legend, calculations need a title, reference number, compiler, checker and continuous page numbering, manufacturer's information needs reference numbering.

9.3.2 Construction design drawings – general issues

- Check for co-ordination clashes with the structure, suspended ceiling and other services and equipment.

- Check the scale and written dimensions for mathematical correctness. Check that the drawn dimensions are referenced from a suitable 'on-site' location.
- Where appropriate check for GMP, food hygiene and other project specific requirements, e.g. materials finish, dust shedding supports, etc.
- Check for routine maintenance access and future plant and equipment removal, e.g. tube withdrawal, filter withdrawal, plant and equipment manufacturers' access requirements, man access into the plant spaces, access for replacement filters, motors or dosing chemicals, abnormal plant and equipment access, cranage requirements, lifting beams, access lifts, clear head height access throughout the space – minimum 2.2 m, access through suspended ceilings to dampers, etc.
- Ensure future provision requirements have been incorporated, spare pump capacity, spare valved connections, speed control on ventilation systems, duty/standby equipment, adequate pipe sizes, etc.
- Check the structural support details with the structural designer and the electrical power and BMS or control information with the electrical designer.
- Ensure the plant, equipment and system noise levels do not exceed the design requirements.

9.3.3 Construction design drawings – schematics

- Carry out a temperature and mass balance on all schematics under varying load conditions
- Review the automatic controls design. Ensure that the provision and location of detectors and control devices are correct
- Ensure that the system can be adequately isolated (note, steam systems require at least two isolating valves between the live steam and the pipework section requiring maintenance), balanced, commissioned and controlled
- Consider system 'what if scenarios', e.g. what happens if the pipework system is over pressurised? What happens to the room pressurisation regime when the filters get dirty?

9.3.4 Construction design drawings – pipework drawings

- Check for adequate provision of air vents, drain cocks, drain valves, isolating valves, regulating valves, commissioning stations, measuring points, air separators, temperature gauges, pressure gauges, flexible connections, depth of gauge pockets, etc.; the installation requirements for this type of equipment should be explicitly detailed in the installation specification
- Check that pipework in running lengths and around installed equipment can be easily removed, e.g. flanges in pipework lengths, unions at coils, etc.
- Check that the space between piped services and between pipes and the structure is adequate for the installation of insulation
- Check that adequate straight lengths of pipe have been provided around commissioning stations and orifice plates
- Check that the pipework system has adequate thermal expansion provision
- On steam systems ensure that the control valve shut-off is slow acting, that all the

proposed valves are consistent with the details in the installation specification, that heating appliances have anti vacuum valves, that all pipes have adequate fall and that strainers are installed on the horizontal

- Check that the thermal insulation specification has been properly understood and applied
- Check that all safety valves terminate in a safe location and that they have been correctly selected for that specific application
- Ensure the system can be easily filled and chemically treated
- Ensure steam and compressed air connections are taken from the top of the supply pipe
- Check the system design and operating pressure is consistent with the design limits of the plant, equipment and system components; confirm the location and method of system pressurisation
- Ensure the cold feed and open vent do not cross the distribution pump set
- Check the length of dead legs and the method of maintaining the system storage and distribution temperatures on hot water systems

9.3.5 Construction design drawings – ductwork drawings

- Check for adequate provision of isolating and balancing dampers, control dampers, attenuators, fire dampers, access panels, access hand holes, constant volume boxes, terminal filters, etc.
- Check a sample of the ductwork velocities particularly around supply diffusers and extract grilles for noise generation
- Check that the termination outlet volumes sum to the fan volume plus a ductwork leakage allowance (see Figure 8.1)
- Ensure AHU equipment has a plinth and is high enough off the floor to allow installation of condensate drains (their depth is at least twice the supply fan developed pressure)
- Check that coil pipework connections are contra flow
- Check that flow measuring detectors have adequate straight lengths of ductwork up and downstream
- Ensure the LEV termination points are less than or equal to their design distance from the point of use
- Check that the range on temperature and pressure gauges matches the actual design requirements, i.e. ensure that the gauge scale can register the measured parameter

9.4 Managing the construction design information

The primary issue with managing the construction design information is to ensure that the information is received, commented on and returned to the contractor *without* affecting the critical path of the construction programme. Although this may appear easy in theory, in practice it can be very difficult to achieve.

The performance or detail design technical specification usually stipulates that the contractor must provide a programme for the submission of construction design information within the first weeks of the contract (usually 2 weeks). It also includes a clause advising the contractor to allow a fixed number of working days for construction design information to be 'checked' by the designer. (Usually 10 days in the receipt of the designer, i.e. excluding any postage period.)

Note: the construction design information submission programme is usually inserted in the quality plan (see Section 10.4) though it may be issued in advance of the plan.

The construction design information is an important part of the contractors services as it:

- Ensures the contractor fully understands the work requirements
- Allows the design aspects of the contractors work to be reviewed
- Reinforces the work quality requirements before installation

However, occasionally the contractor may consider these requirements to be less of a priority than installation progress as this can be the only method used to measure payment. Installing work without agreed construction design information is a recipe for disaster as there is no co-ordination of the work with other trades, no confirmation of quality and the potential for serious project delays if the work needs to be removed later. Before this situation occurs (and in situations where it may not have occurred) it is useful to carry out the following activities:

- Liase with the quantity surveyor before tender issue and ensure that production of the construction design information is integrated into the contractor's payment plan
- Ensure the construction design information submission programme is developed in the first weeks of the project and that it is achievable
- Monitor the contractor against the submission programme; report any programme delays and poor quality of construction design information to the project progress meetings
- Where necessary confirm delays and quality problems in writing to the contractor
- Where the problems persist advise the contractor that lack of improvement will restrict future payments

It is unusual for a project to go this badly wrong although if this action does not create the correct response then the project team will need to consider alternate remedies.

More frequently the opposite scenario occurs. The contractor readily produces the information and swamps the designer who is unable to return the information within the allocated period, potentially leading to a claim for delay. The solution is to provide adequate resources to meet the need in conjunction with a planned submission programme, that includes the likely amount of information submitted. Staff resources are not infinitely flexible or immediately available so advanced planning will be required. Occasionally the project requirements dictate that the designer(s) moves to the site or attends the site specifically to review the information and save the time lost in postage and distribution.

9.5 Chapter review

This chapter explains the design requirements in the construction design information stage of a project through the following topics:

1   Construction design information
2   Content of construction design information
3   Managing construction design information

The construction stage follows the construction design information stage.

Chapter 10

Construction

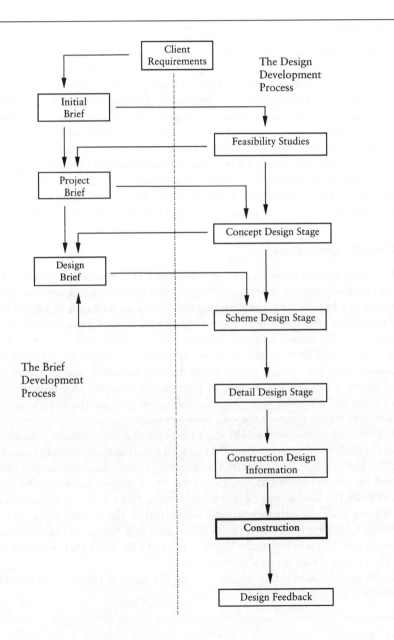

Client
Requirements

The Design
Development
Process

Initial
Brief

Feasibility Studies

Project
Brief

Concept Design Stage

Design
Brief

Scheme Design Stage

The Brief
Development
Process

Detail Design Stage

Construction Design
Information

Construction

Design Feedback

10.1 Chapter contents

The topics covered in this chapter include:

- Construction heath and safety
- Quality plans
- Inspection and testing
- Commissioning and proving
- Handover

10.2 Construction

Although the construction design information and construction activities are separated into distinct stages, in practice they frequently overlap in the need to progress the site-works. This is acceptable from a design viewpoint provided the designer has checked and agreed the construction design information for that section of the work before it starts on-site. However, as site activities transfer the design from paper into physical reality there are a number of project requirements that must be satisfied prior to starting work, and throughout the construction phase. Generally the requirements ensure the work activities are safe and in accordance with the contract conditions. The following sections describe the designer's normal involvement in meeting these requirements.

The emphasis of the designer's role in the construction stage shifts from mostly design verification (in the construction design information stage) to policing.

10.3 Health and safety

In addition to placing duties on the designer the CDM regulations place duties on the client and the individuals or companies appointed as planning supervisor and principal contractor [1]. In summary, the planning supervisor prepares the health and safety file, a document that amongst other things describes the type of hazards and risks involved in the project. (The 'significant' residual hazards identified by the designer in the design CDM reviews will have been passed to the planning supervisor for inclusion in the health and safety file.) At the tender stage the current health and safety file details are passed to the principal contractor as part of the contract documentation. During construction the planning supervisor continues to prepare the health and safety file for final issue to the client at the end of the project.

The principal contractors CDM duties include the development and implementation of the health and safety plan and a range of additional activities intended to ensure a healthy and safe site. The principal contractor is often the main contractor although it could be the management contractor or on a project with significant building services work the lead services contractor. Irrespective of who is appointed as principal contractor all the other contractors involved in the project have specific duties under the CDM regulations that include providing the principal contractor with information to be included in the health and safety file and other activities to compliment the principal contractors' duties.

The designers' CDM duties are essentially the same in the construction stage as in the previous design stages.

- Identify the significant health and safety hazards and risks of any design work
- Give adequate regard to the hierarchy of risk control (see Figure 5.3)
- Provide adequate information on health and safety to those who need it
- Co-operate with the planning supervisor and, where appropriate, other designers involved in the project

Any remaining design work, e.g. design variations or specialist subcontractor work will need to be subjected to the design safety review process established in Section 7.6.

The activities listed below are not legally the responsibility of the designer (strictly they are part of the principal contractors' activities) although they are frequently provided as the designer is often the most informed person to make a judgement:

- Carry out site safety reviews (these will routinely be carried out by the principal contractors' site safety staff, however, reviews from all members of the project team are usually seen as beneficial)
- Review and comment on the building services contractor's method statements.

Method statements describe how the contractor intends to install the works. This includes how the risks associated with the installation process will be controlled and managed. The approach to risk reduction is based on a process similar to the CDM design safety review process, i.e. identify the hazards in carrying out the activity, eliminate as many hazards as practical, highlight the remaining hazards that represent a 'significant' risk, apply the risk control hierarchy and record how the project risks are going to be controlled. Frequently the risks can be controlled by reference to approved codes of practice or manufacturers' instructions, e.g. 'Fibre glass insulation products should be installed in well ventilated spaces. The installer shall only carry out the works with the following PPE in place, a one piece disposable boiler suit (including hood), gloves sealed to boiler suit, eye protection and face mask.'

Method statements will be required for many of the installation activities. It is useful to provide a breakdown structure for the range of statements in the quality plan.

10.4 Quality plan

The quality plan requirements are usually issued as a standard document within the design deliverables package. Most design organisations have their own plan format and content that is tuned to a specific project by adding or deleting standard clauses or ticking a checklist of requirements.

The plan is produced by the contractor. Its aim is to confirm how the contract management and installation quality will be established and maintained throughout the work activities. A plan typically contains [2]:

1   A statement on how the quality plan will be controlled, i.e. the revision and update mechanism
2   Quality auditing procedures
3   Organagrams of the contractors' project structure, clarifying office and site staff and their responsibilities
4   A programme for the works; the initial submission should contain a high level programme with a series of separate work activity programmes behind it that highlight the trade activities and milestones associated with the system or area

of work. Although the level of programming can vary from project to project the final programme must provide enough detail to adequately monitor the work and measure progress (see next section)

5   A method statement breakdown structure, e.g. general activities, pipework installation, insulation installation, etc.
6   The construction design information submission programme
7   A pack of standard forms, e.g. requests for information, test certificates, inspection certificates, etc.
8   Procedures for all the main activities including:
    –   Procurement
    –   Off-site manufacturing, inspection and testing
    –   Packaging, delivery and storage on-site
    –   Protection of the works
    –   Requests for information (see Section 10.5)
    –   Site supervision
    –   Site inspection
    –   Testing
    –   Commissioning
    –   Progress payments (see Section 10.5.4)
    –   Handover

The quality plan will remain a 'live' document throughout the construction period with all the subcontractors and specialist subcontractors adding to it.

The designer needs to review and check the information presented in the plan to ensure that it is consistent with the requirements specified in the design deliverables.

10.4.1 Programming issues

Although the designer is not directly responsible for programming the building services contractor's work, the designer will need to comment on the programme integrity. The designer will also want to ensure the programme is adequately detailed to realistically monitor progress and establish a method for cost measurement. Although programming software has improved the contractor's ability to prepare and present installation programmes it still takes a considerable effort to produce a realistic programme. It is also worth bearing in mind that the programme is rarely correct as it is an estimate of future work that is influenced by a series of issues the programmer cannot control. However, awareness of typical programme problems can help reduce programme risk. Typical issues include:

*   Lack of integration with the main project programme; frequently the building services contractor is introduced to the project after the main contractor is appointed; if the main contractors' programme does not form part of the building services tender package or the imposed time scales are unrealistic the two programmes may not be able to be integrated without significant changes
*   Inadequate time for commissioning; the commissioning specialist will provide the best programme estimate, however, where this is not possible the designer and the contractor will need to detail the commissioning requirements and establish a suitable time scale

- Ensure that adequate notice is provided to all third party inspectors and that their 'sign-off' and inspection visits are integrated into the programme, e.g. the fire officer, the EHO, FDA inspectors, etc.
- Ensure the utility services are installed, commissioned and signed off prior to any commissioning work. Otherwise there may be no electrical power, water, gas, drains, etc. and consequently probably no commissioning!
- Ensure the mechanical and electrical services programmes are integrated as the mechanical services will not work without electrical power
- Ensure the building fabric 'weather tight' milestones are shown on the building services programme as the majority of the installation work cannot be carried out without a weather proof environment
- Ensure adequate time is allocated to client training
- Ensure the operation and maintenance documentation is prepared, reviewed and revised in advance of the handover date; handover can be delayed if the client has incomplete or inadequate information to safely operate the facility
- Ensure that all specialist subcontractor work including design is identified in the building services programme
- Ensure the work breakdown is adequately detailed and allows realistic progress monitoring

10.5 Inspection and testing

The previous construction stage activities should be considered as pre-installation work as they describe how the work will be done. Once the work starts on-site there is a need for physical inspection and testing of the 'on going' work. Frequently these activities are delegated to a resident building services clerk of work. (Where the project requirements split the designers role across a number of different people it is very useful to clarify the individuals roles to the project team to prevent any misunderstanding.)

10.5.1 Testing

The designer or clerk of work will apply the testing criteria defined in the installation specification through the agreed testing procedure established in the quality plan. The range of test requirements is considerable and includes pressure testing, safety testing, leakage testing, load testing, integrity testing, noise and vibration testing, etc. All major plant items and systems may need to be tested (on or off site) e.g., boiler plant, compressors, air handling units, pumps, attenuators, control panels, smoke ventilators, fans, installed pipework systems, installed ductwork systems, etc. Subject to a successful outcome the designer or clerk of work will issue a test certificate to confirm that the component or section of the system has met the performance requirements. The aim is to confirm that the individual system components meet the design requirements before they are integrated into the complete system.

10.5.2 Inspection

Inspection requirements tend to consider the quality of the installed work and are generally operated as a passive check, as opposed to active testing. The installation requirements are detailed in the construction design information and the installation specification. The inspection process involves the designer or clerk of works walking the site and checking the installation against these documents. Each site inspection is recorded in a site inspection report that includes details of any inconsistencies or snags that need to be actioned before that section of the work can be signed off. (Typically snagging issues are similar to the review issues raised in Sections 9.3.4 and 9.3.5.)

The inspection process also provides the designer or clerk of work with a good understanding of the contractor's site progress. Progress meetings are usually held on-site weekly or fortnightly. These provide a mechanism for progress reporting and where necessary agreeing actions to resolve current progress problems. The designer usually attends these meetings to ensure the design requirements are fully understood, to report progress on their review of construction design information, RFIs, etc. and action any current building services issues, ideally before they become critical.

10.5.3 Requests for information

During the site works it is likely the contractor will have a number of queries on the design and installation work. The normal process for recording queries is through a request for information (RFI) sheet. Like the construction design information, these usually need to be returned speedily to the contractor to prevent programme delays. It is difficult to plan for RFIs as they are obviously unplanned events. Fortunately the answers are often already in the design deliverable information and few usually require significant effort. However, the RFI format, recording procedure, details of the contact staff, etc. will need to be agreed and included in the quality plan.

Occasionally RFIs or the actual installation work will raise physical space constraints that were not obvious during earlier stages in the design process, e.g. access to the rear of a flange is blocked by a cable tray. These issues are generally best resolved on-site between the appropriate contractors and clerk of works. However, a technical variation order may need to be issued to confirm the way forward and where necessary provide a mechanism for the contractor to recover any costs.

Provided the design development has followed the process described in the earlier chapters, i.e. value management, design brief, value engineering, specification content and checking of the contractors' design information it is unlikely that significant technical variations will occur. In the event of this kind of change there will be an auditable trail to demonstrate that the designer has completed the design in a professional and competent manner. All design work carries a technical risk and while the design development process minimises that risk, it cannot eliminate it. Consequently earlier discussion with the quantity surveyor during the preparation of the contract documents should include the allocation of a risk budget to the building services contractors tender cost (see Section 7.9.1). The size of the risk budget will be determined by a number of factors including quality of the tender design deliverables, the contractors knowledge and understanding of the works, the spread of the bid costs, the form of contract, etc. Without knowledge of the specific project criteria it is difficult to define a correct figure, however, previous experience suggests, as a guide that a technical risk budget of 5 per cent of the building services cost of work will prove adequate.

Note: the construction management risk issues are not included in this guidance figure and will need to be separately assessed.

10.5.4 Progress payments

The conditions of engagement generally require designers to provide 'technical advice' on certificates for contractor payments made by the quantity surveyor. The advice usually consists of an agreement between the designer and the contractor on the month's installation progress and a payment figure for the new work.

The methodology for agreeing payment figures is not generally written into the contract documents. Industry practice usually breaks down the contract sum into sections for pipework, ductwork, etc. and then distributes them again into subsystems such as water services pipework, LTHW pipework systems, chilled water systems, etc. Once a value has been allocated to the systems, the designer and contractor walk the site and agree a 'percentage complete' figure for that activity. The total figure is then calculated by multiplying the percentage complete against each activity value and summing all the activity figures. The monthly payment is then determined by subtracting the sum of the previous monthly figures from the total figure, subtracting a retention figure (usually 5 per cent) and what remains is the contractors monthly payment.

This can become a very elaborate process that is awkward to justify and easily complicated by the contractor asking for payment for materials stored off-site or stored on-site but not installed. Payment issues should be discussed with the quantity surveyor before the contract is let. If the surveyor has a preferred method for determining payment this may need to be applied otherwise the pragmatic solution is to define the payment methodology in the quality plan requirements and agree it with the contractor before the contract is signed.

The quality plan requirements should instruct the contractor to provide a breakdown of the contract sum by the main constituent parts, e.g. pipework, ductwork, etc. for each main area and then into subsystems, e.g. water services, LTHW pipework systems, etc. listing the build-up of labour and materials in each subcategory. (Most contractors have software estimating packages that automatically provide this level of detail as they need it to accurately bid the work.) The cost breakdown structure should mirror the structure of the main programme activities. Although it is likely that each of these programme sections will be broken down into a number of activities, e.g. first fix, second fix, etc. it does allow a correlation to be made between the programmed works, the actual percentage complete and the value of the activities.

Where plant and equipment items need to be procured in advance of the site works they should be allocated as a separate cost sum within the contractor's breakdown. This can be drawn when the goods have been registered as owned by the client. Alternatively the client can buy the goods directly and free issue them to the contractor though the co-ordination aspects of this approach will require careful examination.

10.6 Commissioning

10.6.1 Commissioning activities

The majority of the designers activities need to be completed significantly in advance of any commissioning work. The activities include:

- During the design period ensure the systems can be commissioned (Section 8.5).
- Ensure the extent of the system commissioning requirements and the commissioning standards are clearly detailed in the technical specification. The requirements are usually defined in the installation specification. The commissioning procedures described in the standards should be incorporated into the contractors quality plan. *Note*: the chapter reference and further reading sections include the common commissioning standards.
- Confirm in the contract documents the requirement for an independent commissioning specialist and the specialist's role. The role should include co-ordination of all commissioning requirements including those packages of work that maybe outside the building services contractors scope of work, e.g. fire protection systems, BMS, etc. (The role is usually defined in the installation specification.)
- Provide all the design information necessary to successfully commission the systems. This could include:
  - Commissioning tolerances
  - System and design data including operating pressure, temperature, humidity, noise rating, room differential pressure, occupancy periods, acceptance criteria, etc.
  - Construction design information, including general layout drawings detailing pipe sizes, duct sizes, access doors, dampers, valves, etc.
  - Schematic drawings indicating the design intent
  - Pump and fan performance curves
  - Plant and equipment schedules cross-referenced to the schematics, confirming sizes, flow rates, pressure drops, valve authorities, throws, etc.
  - Plant and equipment wiring diagrams
  - Controls wiring diagrams and operation description
  - Cause and effect charts, e.g. when an event occurs what are the system reactions
  - A comprehensive system, plant and equipment labelling system
  - System proving requirements

  Note: some of this information will be provided by the contractor as construction design information.

- Ensure commissioning activities are planned as early as possible by the commissioning specialist and integrated into the main project programme. Commissioning invariably takes more time than is initially allocated in the project programme as the full implications require thorough investigation of all the work packages and the contract documentation requirements.
- Confirm the requirements for statutory, regulatory and insurance approvals.
- Ensure the pre-commissioning activities have been successfully completed (see next section).

- Provide support to the commissioning staff during commissioning and witness the results.

From the designer's point of view successful commissioning is a prerequisite for successful system proving, so it is to the designer's benefit to be proactive during the commissioning works.

10.6.2 Pre-commissioning checks

Prior to the commissioning of the systems the following pre-commissioning checks should be successfully completed:

- Pipework pressure testing or ductwork leakage testing (where necessary) for the complete systems should be confirmed and certified
- Any snagging issues generated by the site inspections should be completed
- Final inspection checklists for the systems should be signed off
- Pipework systems should be flushed, cleaned, filled, treated and vented
- Ductwork systems should be cleaned
- The area around the system should be clean and allow easy access for commissioning activities
- The utility services supplies to the system should be commissioned and safe to use; any appropriate safety systems should also be enabled

Pre-commissioning checks need to be signed off by the clerk of works or designer to confirm the system is ready for commissioning. A comprehensive range of commissioning check lists is provided in Refs. [3,4].

10.6.3 Commissioning

The commissioning process for pipework and ductwork services involves the following steps:

- Setting of the system to work; the distribution system is 'set up' for regulation and the correct operation of the main plant items is checked
- Regulating the system; the distribution system is proportionally balanced and then regulated to supply the design requirements
- Plant performance testing; the operation of the plant is tested to ensure it operates in accordance with the design requirements

After completion of each stage in the process the contractor offers the system to the designer for witnessing and subject to it being acceptable it is 'signed off'.

10.6.4 Design and acceptance criteria proving

This is a defining period for the designer as it represents the ultimate test of the design work. Invariably the designer will want to be on-site during the proving tests.

After the commissioning tests have been 'signed off' the systems are run to prove that they can maintain the design performance across the range of test criteria. Any acceptance criteria that relate to the design performance will be integrated into the design proving tests.

Proving considers system control, system operation in differing conditions and system capacity. Typically this may consist of:

- Independently confirming the accuracy of the control detectors, e.g. independently monitoring space conditions to ensure the detectors are correctly calibrated
- Confirming the BMS or control system is operating correctly, e.g. detection signals are correctly fed back to the control system, correctly interpreted and correctly sent out to the controlling device
- Confirming the control device physically operates in conjunction with the control signal, e.g. the valve response is proportional to the signal instruction
- Confirming that the space conditions are maintained within the design tolerance for a defined period of time in normal use; 24 hour logs are usually available from the BMS system
- Confirmation that the integrated systems operate correctly in normal, standby and emergency modes; some of these tests will need to be repeated to external third parties certifying the facility, e.g. fire officer, EHO, etc.
- Confirming performance of systems that may not be directly linked to the BMS or control system, e.g.
  - Sprinkler flow test
  - Hydrant flow and pressure tests
  - Domestic water service flow, pressure and time to temperature tests
  - Clean room particle count tests
  - Environmental, space and equipment noise level checks
  - Plant, equipment and system 'touch' temperature tests
  - Specialist gases detection systems
  - Occupation exposure levels (OELs), e.g. underground park CO_2 levels
  - Critical pressure regimes between spaces, e.g. hazardous and non-hazardous areas
- Confirming the systems capacity; simple capacity failures will become apparent during the control tests, however, specific tests may need to be designed to simulate unique design conditions; load variations created by seasonal changes are usually monitored over a 12 month period; certain types of proving tests will need to be carried out when the facility is occupied, e.g. OELs, air conditioning space loads, fine tuning of the system, etc.; these will need to be identified separately from the tests carried out during the construction period and where possible, integrated within the clients induction period

10.7 Handover

After successful system proving and prior to handover of the facility the designer will need to ensure that the following activities are closed out.

- Final inspections. The majority of defects should have been actioned as part of the pre-commissioning procedures. Final inspections will highlight any minor defects or snags that were not previously apparent. Providing these snags do not impair the safe and reliable operation of the systems they 'may' be added to a final snag list and issued with the practical completion. These will need to be satisfactorily completed during the defects liability period (usually 12 months from practical completion) to enable the contractor's retention money to be released. It is usually beneficial for the designer or clerk of work to carry out the final inspection with the contractor so that snags are agreed and the level of re-work understood.

- Operation and maintenance documents. The facility cannot be handed to the client unless they 'are provided with adequate information about the use for which the article is designed' (HASAWA 1974). The building services information also forms an important part of the health and safety file. The format and proposed contents of the O&M documents are outlined in the design deliverables. As client requirements for O&M information continue to increase most design organisations specify a specialist independent company to formulate the documentation. The designer needs to ensure that these requirements have been satisfied in advance of the handover date. This can be problematic as commissioning and proving tests may not be complete until virtually the end of the construction period. The pragmatic solution is to review the draft information at least 7–8 weeks beforehand, review the final document (except the final commissioning and proving information, if this is not available) 3 or 4 weeks later and insert the final test certificates a week before handover. The work needs to be carefully planned and integrated with the planning supervisors requirements, otherwise the documentation will be incomplete and handover may be delayed.
- Keys, tools and spares. The design deliverable requirements should be checked and the equipment issued to the client.
- Certification and facility sign-off by the fire officer, EHO, clients insurer, etc.
- Client training. Many building services systems are complex and require a considerable period of training. Training requirements will have been specified in the design deliverables, however, it is frequently beneficial to supplement the client training with their involvement in the commissioning and proving work. In certain circumstances the contractor and or control specialist may need to retain a presence on-site during the client induction period and after handover to ensure a smooth transition of services.

The building services aspects of the project will be successfully handed over if these issues are closed out before the handover meeting.

10.8 Chapter review

This chapter explains the building services construction requirements through the following topics:

1   Health and safety
2   The quality plan
    - Programming issues
3   Inspection and testing
    - Progress payments
4   Commissioning
    - Commissioning activities
    - Pre-commissioning checks

- Commissioning and proving
- Design and acceptance criteria testing

5   Handover

References

[1] *A Guide to Managing Health and Safety in Construction*, HSE Books.
[2] *Project Management Handbook for Building Services*, Building Services Research and Information Association Publications, Application Guide AG 11/98.
[3] *The Commissioning of Water Systems in Buildings*, Building Services Research and Information Association Publications, 1998, Application Guide AG 2/89.2.
[4] *The Commissioning of Air Systems in Buildings*, Building Services Research and Information Association Publications, 1998, Application Guide 3/89.2.

Further reading

Wild, L.J., *Site Management of Building Services Contractors*, E&FN Spon, London, 1997.
Pre Commissioning Cleaning of Water Systems, Building Services Research and Information Association Publications, Application Guide AG 8/91.
Water Treatment in Building Services Systems, Building Services Research and Information Association Publications, Application Guide AG 2/93.
The Chartered Institution of Building Services Engineers, *Commissioning Codes*, CIBSE Publications: Code A, Air distribution systems (1996); Code B, Boiler plant (1975); Code C, Automatic controls (1973); Code R, Refrigeration systems (1991); Code W, Water distribution systems (1994).
Handover Information for Building Services, Building Services Research and Information Association Publications, Technical Note TN 15/95.

Chapter 11

Design feedback

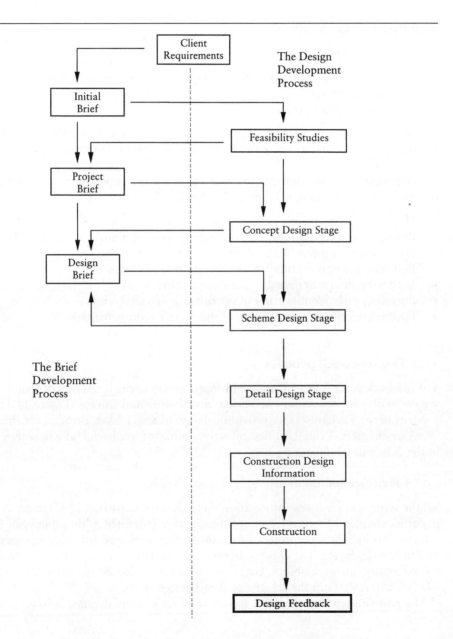

11.1 Chapter contents

The topics covered in this chapter are:

- Reasons for feedback
- The feedback process
- The design process summary and feedback

Detailed information on building services design feedback is provided in Feedback for Better Building Services Design, BSRIA Application Guide AG 21/98.

11.2 Reasons for feedback

The fundamental reasons for feedback are [1]:

- To improve the current project (i.e. improve the project match with the client's key success criteria)
- To improve future projects

Feedback essentially highlights the issues that are successful and those that require improvement or rejection and ensures they are incorporated into current and future project work.

The primary feedback benefits to the designer are the opportunity to reduce design risk and improve the design solution. Feedback achieves this by:

- Highlighting opportunities to improve the brief and design development process
- Providing informed design decision making and better understanding of the design impact on the project
- Reducing the opportunity for site variations and cost extras
- Improving design knowledge and design deliverables
- Providing early identification of potential design problems
- Ensuring a better understanding of the client's requirements

11.3 The feedback process

The feedback process is a virtuous self-improvement cycle. It consists of four separate stages: feedback information, definition, distribution and storage (Figure 11.1).

Many of the mechanisms for providing design feedback have already been discussed in earlier chapters. Table 11.1 describes the sources of feedback and where they occur in the design development process.

11.3.1 Feedback information

All the sources of feedback information up to the post-construction stage are detailed in earlier chapters. From the concept design stage (Chapter 5) the sources of design information include past project work, other engineers, end users, design guidance, British Standards, etc. The concept option evaluation review considers constructability, reliability, maintainability, etc. These issues may also be the subject of separate independent reviews in the scheme or detail design stage.

The post-construction feedback information sources are detailed below.

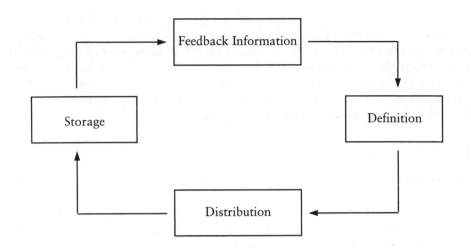

Figure 11.1 The feedback process.

11.3.1.1 Internal post project review

Internal post-project reviews are an opportunity to consider global project issues and capture design feedback on issues such as design deliverables that do not form part of the other reviews.

Typically the review is chaired by a person who has had little involvement in the project and acts as facilitator to clarify and draw out information. The lead designers

Table 11.1 Sources of feedback information in the design development process

Design stage	Feedback information
Feasibility studies	Project evaluation and value management review
Concept design stage	Review of the sources of design information
	Option evaluation review
	CDM design review
Scheme design stage	Value engineering report
	CDM design review
	Technical design review
Detail design stage	Good practice feedback review
	Value engineering report
	CDM design review
	Technical design review
Construction design information	Construction design information review
Construction	Contractors' method statements
	Contractors' quality plans
	Project programming issues
	Inspection and testing reports
	RFI and variation files
	Commissioning reports
	Final inspection reports
Post-construction	Internal post project review
	Client feedback report
	Energy audit

are asked to present six items on the project that went well and six that could be improved. (The number of items should be limited to ensure the key issues are flushed out although this guideline needs to be flexible enough to allow other obvious issues to be presented.) The issues can be anything to do with the project, e.g. design management, administration, resourcing, technical solutions, etc. The results are summarised by the facilitator onto project feedback sheets, circulated to the designers and copies placed in the central feedback information file (see Section 11.3.2).

11.3.1.2 Client feedback report

All comprehensive quality systems involve client feedback as a mechanism to review current performance and improve the product or service. The design process is the primary service that design organisations provide to their clients and consequently feedback is invaluable. Often feedback is based on informal discussions or comments from the client. Unfortunately these frequently focus on the 'current' problems and although of value do not provide a structured, repeatable basis for monitoring service improvement. A better option is to prepare a feedback questionnaire or comment sheet that the design organisations representative can use as a basis for a formal client feedback discussion. The content of the questionnaire would include broad service issues such as fee, service quality, ability to respond, future use recommendation, etc. as well as technical design issues.

In certain circumstances direct feedback from the end users may be available through a post-occupancy survey. This type of feedback is very useful although often difficult to obtain on a formal basis.

11.3.1.3 Energy audit

In certain circumstances the maximum energy consumption of the facility forms part of the clients contractual requirements or one of the client's key success factors. Post-occupancy monitoring will be required to confirm the facility performance and the success of the design predictions.

11.3.2 Definition

Over the duration of the project the feedback information builds into an impressive quantity of paper. Ideally copies of each set of information will be lodged in a dedicated 'feedback information' set of files that grow as the project develops. After project completion or in stages during the project, the information needs to be reviewed, summarised and the key lessons defined into easily referenced and usable information.

A simple approach is to prepare a broad list of main subject categories and a standard feedback sheet. The categories need to be broad enough to apply to all types of projects, e.g.

- Design management, e.g. resources, design team structure, use of information technology, design programming, quality of the design brief, design co-ordination, etc.
- Technical design (subdivided into different disciplines), e.g. use of chilled beams,

natural ventilation feedback, BMS developments, innovative design solutions, good suppliers, etc.
- Project issues, e.g. operation of the project structure, client end user feedback, successful business risk control, integration with other design and project team members, etc.

The standard feedback sheet needs to convey enough information to allow a designer to quickly understand the issue, provide a route to more detailed information and allow easy referencing. A sample feedback sheet is described in Figure 11.2. Any action required as a consequence of the feedback needs to be reviewed by the design manager and where appropriate incorporated into the management procedures.

The feedback sheet (and any simple attachments, i.e. a letter or table) should be referenced and placed in the design organisations central feedback information file.

Most design information is produced on software or can easily be scanned into a digital format. Consequently there is no need to place lots of paper copies of documents in the central feedback file, only software file references.

11.3.3 Distribution

Feedback information generated during the project will be circulated throughout the project team via the project distribution system. This is likely to be as a paper copy or if there is a digital document management system via a posting on the system.

The internal distribution of post-construction feedback should include all the lead designers involved in the project and where appropriate the other lead designers in the organisation. The project feedback sheets need to be distributed to all the designers so that everybody understands the lessons learnt. Distribution of paper copies may be acceptable if the number of designers is relatively small otherwise the simplest solution is to set up a feedback bulletin board on the e-mail system.

11.3.4 Storage

The central feedback file should be held in the technical library or other similar location to allow easy access for designers.

It is common practice to physically archive project documents when the project has expired. Frequently the archived boxes of documents are not readily accessible (e.g. 24 hour notice) and they absorb lots of space. Consequently it makes sense to digitally archive as much information as possible. Efficient project software directories set up in advance or early in the project life can easily be archived and require little physical space. The file references in the feedback sheets allow designers quick and easy access to the original files for fuller information and provide a source of material for new design work.

Feedback is the last stage in the current design project and the first stage in the next.

11.4 The design process summary

The design process is fairly complex and essentially a 'people' process. The book may outline the process requirements, however, the designer will implement them. The

Project Feedback Sheet	
Date :	Sheet Reference No :
Project No :	Archive Box No :
Category :	Staff Contact :
Software File Reference (of related documents) :	
Subject	
Key Learning Points	
Action Required	

Figure 11.2 A standard feedback sheet.

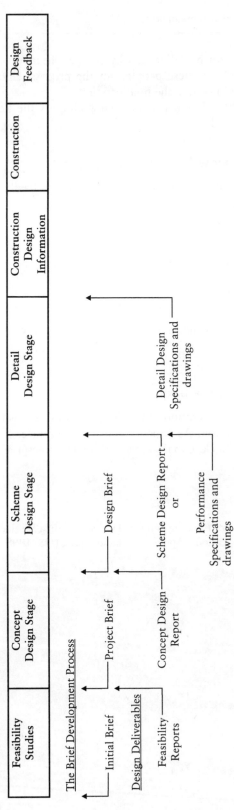

Figure 11.3 A summary of the building services design process.

designer, in turn, will be influenced by the people involved in the project and the project will be built by these people, not the process. Understanding and applying the process simply improves the final product.

A summary of the design process that highlights the key issues at each stage is described in Figure 11.3.

11.5 Chapter review

This chapter explains the requirements of the design feedback through the following topics:

1   Reasons for feedback
2   The feedback process
    • Feedback information
    • Definition
    • Distribution
    • Storage
3   The design process summary

This is the last stage in the design process.

11.6 Feedback on the book

This work is my interpretation of the design process and is based on research and my experience and opinion. As I mentioned in the preface, it is a starting point and I welcome your feedback, e-mail: david@abownass.freeserve.co.uk

References

[1] *Feedback for Better Building Services Design*, Building Services Research and Information Association Publications. Application Guide AG 21/98, 1998.

Index